T0400385

Vortex Formation in the Cardiovascular System

Arash Kheradvar • Gianni Pedrizzetti

Vortex Formation in the Cardiovascular System

Authors
Arash Kheradvar, M.D., Ph.D.
University of California, Irvine
2410 Engineering Hall
The Edwards Lifesciences Center for
Advanced Cardiovascular Technology
Irvine
California 92697
USA

Gianni Pedrizzetti, Ph.D.
Università degli Studi di Trieste
Dipto. Ingegneria Civile e
Architettura
P. Europa 1
34127 Trieste
Italy

ISBN 978-1-4471-2287-6 e-ISBN 978-1-4471-2288-3
DOI 10.1007/978-1-4471-2288-3
Springer London Dordrecht Heidelberg New York

British Library Cataloguing in Publication Data
A catalogue record for this book is available from the British Library

Library of Congress Control Number: 2011944219

Printed on acid-free paper

Springer is part of Springer Science+Business Media (www.springer.com)

To my wife Ladan and our children Aryana and Ario for their love and support

Arash Kheradvar

To my wife Ilaria and our children Giulia, Guglielmo, Greta for accompanying with smiles and infinite love.

Gianni Pedrizzetti

Preface

The topic of vortex formation has received much attention over the past few years. Vortices occur in nature wherever propulsive flow exists; from erupting volcanoes to the ones generated by squid and jellyfish to propel them. Vortices drive the atmospheric circulation; smaller ones like hurricanes and tornado contain an enormous amount of destructive energy. They are also present in the heart.

A wide variety of vortices develop in the cardiovascular system, particularly in the cardiac chambers, and in large arteries. These vortices play fundamental roles in the normal physiology and provide proper balance between blood motion and stresses on the surrounding tissues. In contrast, formation of unnatural vortices may alter the momentum transfer in the blood flow and increase energy dissipation.

In the present book, we have tried to recapitulate the current knowledge on the vortex formation in the cardiovascular system, from physics to physiology. The first two chapters of the book cover the fundamental aspects of fluid mechanics and in particular vortex dynamics. Through these, the reader's attention is driven toward the lifespan of vortices – from formation to dissipation and interaction. The third chapter describes vortex formation in heart. Through this chapters, formation of vortices at different locations inside the heart, and their physiological and clinical significances are discussed. Chapter four discusses the effects of cardiac devices and surgery on vortex formation. In a nutshell, the fluid dynamics of the artificial heart valves and ventricular assist devices are described, and the effect of each device on cardiovascular vortex formation is comprehensively reviewed. The last chapter is focused on diagnostic vortex imaging in which we have reviewed existing methods to visualize vortices in the cardiac chambers, and the experimental and computational techniques to model the vortex formation in cardiovascular system.

The material described in the book has been brought together from the published and unpublished work of the authors and the invited contributors. Through this work, we desire to translate physical, mathematical and engineering concepts related to vortex formation into a clinical perspective with the objectives that these notions finally advance the cardiovascular patient care.

This work is the product of three years of transatlantic collaboration on various cardiovascular projects between Kheradvar's and Pedrizzetti's groups. In addition, we owe debt of gratitude to the invited contributors who permitted to span over the multidisciplinary expertise involved in this field; without their support and involvement, completion of the book seemed impossible. We would also like to acknowledge our students and postdoctoral scholars, Ahmad Falahatpisheh, Hamed Alavi, Jan Mangual, Brandon Dueitt, and our colleague Federico Domenichini who helped us preparing the figures, plots, etc. At the end, we would like to acknowledge the American Heart Association for providing funding support for the right ventricular vortex project (10BGIA4170011). The funding deeply helped us contribute to understanding of vortex formation in the right heart and enhanced our collaboration that gave rise to the present book.

Arash Kheradvar, M.D., Ph.D. — Irvine, CA, USA
Gianni Pedrizzetti, Ph.D. — Trieste, Italy

Acknowledgements

The authors would like to thank the below authors for their contributions to this book:
Haruhiko Abe, M.D., Chapter 3: Sects. 3.5 and 3.6.1, 3.6.2, 3.6.3.
Osaka Minami Medical Center, Japan

Idit Avrahami, Ph.D., Chapter 4: Sects. 4.2 and 4.3. Chapter 5: Sect. 5.4.
Ariel University Center of Samaria, and Afeka Academic College of Engineering, Tel Aviv, Israel

Giuseppe Caracciolo, M.D., Chapter 3: Sects. 3.5 and 3.6.1, 3.6.2, 3.6.3.
Mount Sinai School of Medicine, New York, NY, USA

Gabriele Dubini, Ph.D., Chapter 4: Sect. 4.4.
Politecnico di Milano, Milano, Italy

Tino Ebbers, Ph.D., Chapter 5: Sect. 5.1
Linköping University, Linköping, Sweden

Jan Engval, Ph.D., Chapter 5: Sect. 5.1
Linköping University, Linköping, Sweden

Kambiz Ghafourian, M.D., Chapter 3: Sects. 3.6.4. and 3.6.5.
Washington Hospital Center, Washington, D.C., USA

Helene Houle. Chapter 5: Sect. 5.2.
Siemens Medical Solutions, Mountain View, CA, USA

Francesco Migliavacca, Ph.D., Chapter 4: Sect. 4.4.
Politecnico di Milano, Milano, Italy

Laura Miller, Ph.D., Chapter 3: Sect. 3.4.
University of North Carolina, Chapel Hill, NC, USA

Partho P. Sengupta, M.D., Chapter 3: Sects. 3.5 and 3.6.1, 3.6.2, 3.6.3.
Mount Sinai School of Medicine, New York, NY, USA

Contents

Chapter 1
Fundamental Fluid Mechanics

Abstract This chapter of the book introduces the basic elements of fluid mechanics constituting the essential background for understanding the blood flow phenomena in the cardiovascular system. It discusses the physics of flow and its implications. This chapter is aimed to provide an intuitive understanding, accompanied by an essential mathematical formulation that ensures a rigorous reference ground.

1.1 Fluids and Solids, Blood and Tissues

The most definitive property of fluids, which include liquids and gases, is that a fluid does not have preferred shape. A fluid takes the shape of its container regardless of any geometry it had previously. In contrast, a solid consists of constituting elements with a predefined shape. When the relative position of these constituent elements is infinitesimally changed, internal stresses develop to restore the elements to their original, stress-free state. This distinctive property of solids is called elasticity. An elastic deformation typically is completely reversible as the energy stored in the deformed elements is totally released when the deformation ceases.

Fluids, on the other hand, do not share this feature of the solid materials. They have no preferred geometry; thus they possess infinitely independent, stress-free states. Nevertheless, fluids exhibit an internal resistance during their relative motion. This resistance is due to the development of internal stresses in response to a "rate of deformation". This behavior is due to *viscosity*. Therefore, a fluid experiences a viscous resistance during the motion, which is caused due to sliding fluid elements on each other. Given that the viscous stresses represent a frictional phenomenon that appears during motion, when the motion is ceased, no internal stress returns the fluid to its original state, as in the solids. The mechanical energy that deforms the fluid elements is not being stored anywhere; it dissipates due to internal viscous friction, which is transformed into heat and

A. Kheradvar, G. Pedrizzetti, *Vortex Formation in the Cardiovascular System*,
DOI 10.1007/978-1-4471-2288-3_1, © Springer-Verlag London Limited 2012

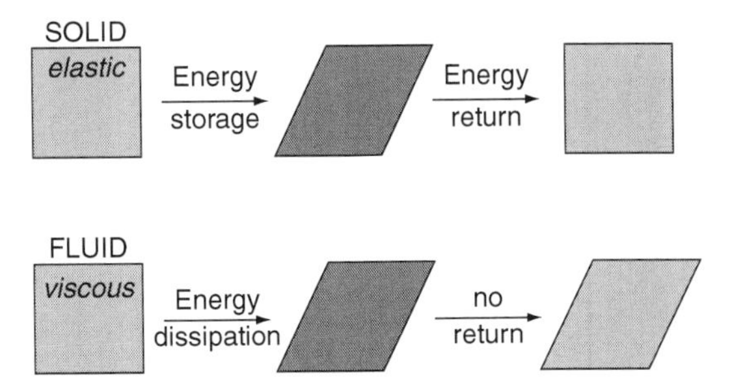

Fig. 1.1 Solid materials are characterized based on their elastic behavior. Due to elastic deformation of materials, the elastic potential energy is stored in the elements of that solid, which is released when the solid returns back to its original shape. Alternatively, the viscous behavior of fluids appears as an internal resistance during the deformation process, which is due to internal shear-stresses that are friction-driven, and are associated with the dissipation of energy

dispersed away. This energetic difference between elastic and viscous behaviors is sketched in Fig. 1.1. However, the distinction between fluids and solids is not as sharp. Most materials present both elastic and viscous behaviors. Some materials can behave either as fluids or solids in some respects. For example, a glacier is a solid if one can walk on it, yet it flows like a fluid during its slow motion over the years.

Blood is composed of deformable cells (elastic elements) immersed into plasma (fluid element). Therefore, blood is not a simple material; rather it is a mixture of heterogeneous elements. If the dimension of the cells is comparable with the size of the container, the corpuscular nature of the blood takes a fundamental role in the physical processes occurring at such a scale. One example is the blood flow in the capillaries where red blood cells (RBCs) as biconcave disks with a diameter of about 7–8 μm must deform to pass through the vessels with diameters as small as 5 μm. Flow in the arterioles as well as venules is also directly influenced by the corpuscular nature of blood. As the diameter of the blood vessel increases, the influence of individual RBC progressively decreases. Every 1 mm^3 of blood contains about 2 million RBCs. Therefore, it is estimated that blood flow in vessels with diameters larger than 1 mm is rather continuous than granular. This representation of blood allows employing a rich theoretical background of continuum mechanics and differential mathematics to solve problems involving biological flows.

Once assuming blood as a continuum, its corpuscular nature is represented by viscosity, which cannot be considered constant. In fact, the apparent blood viscosity is not an intrinsic material property, and thus changes its value depending on the type of blood motion at different sites. For example, blood viscosity is reduced in regions with high shear rates when the blood cells are separated away and the observed friction is mostly due to the plasma. Conversely, the viscosity is increased in the central part of a rotating duct due to denser population of RBCs there. Therefore, as a general rule, the viscosity of blood is a function of the percent concentration of RBCs in blood or local hematocrit. Such variability is influenced by several factors, and is usually small. However, evaluation of such small variations is difficult particularly for three-dimensional flows with whirling motion. Therefore, flow in large vessels is usually treated as a Newtonian fluid, which is a continuous fluid with constant viscosity whose value is about three times greater than the viscosity of water.

Occasionally, when friction is negligible compared to other existing factors, blood behavior may be approximated as an ideal fluid with no viscosity. Such approximation is the basis of the Bernoulli theorem (to be discussed in Sect. 1.4). The Bernoulli theorem is useful in computations involving brief tracts or situations where blood elements are away from the vessels' boundaries. However, viscous forces are never negligible adjacent to the solid boundaries.

1.2 Conservation of Mass

The first physical law governing the mechanics of blood as a continuum is the *conservation of mass*, or *law of continuity*. The conservation of mass states that the difference between the flow that enters and leaves a certain container is equal to the variation of the volume of fluid in that container. In general, this principle also accounts for the variation of the fluid density due to either compression or dilation. Blood is essentially incompressible under physiological conditions, meaning that its density cannot vary appreciably.

When applied to systems with rigid walls, continuity states that the flow that enters through a rigid vessel is identical to the flow that exits at the same instant. In other terms, the discharge flow inside a rigid vessel is the same when measured at any cross-section of the vessel, independent of the vessel's geometry. The discharge or flow-rate, Q, is given by the product of the area of the cross-section, A, and the blood velocity herein, U,

$$Q = U \times A; \tag{1.1}$$

where, U is the longitudinal velocity averaged over the whole cross-section. The continuity law states that Q is constant along a rigid vessel. Therefore, if the cross-sectional area A decreases, the velocity U should necessarily increase to keep their product constant. As a result, the flow moves faster where the diameter is less and slower where it is more.

The concept of continuity is also valid for the flow entering into a compliant chamber. If the chamber volume, V, varies during time, the entering flow-rate, Q_{in}, is not necessarily equal to the exiting one, Q_{out}, and their difference corresponds to the fluid stored in the compliant chamber

$$Q_{in} - Q_{out} = \frac{dV}{dt} \tag{1.2}$$

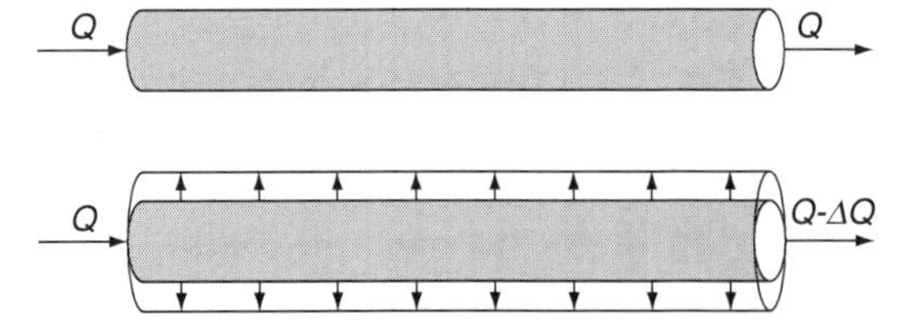

Fig. 1.2 In a rigid vessel (*above*), the flow-rate cannot change along the vessel, in an elastic vessel the flow-rate is reduced downstream during the vessel expansion because part of the incoming fluid is stored laterally

As an example, for the case of the left ventricle, during systole when the mitral valve is closed and $Q_{in}=0$, the difference between end-diastolic and end-systolic volumes corresponds to the volume flown into aorta. Similarly, the same volume variation corresponds to the transmitral flow during diastolic filling.

A compliant vessel is able to store part of the incoming fluid during expansion. Therefore, the discharge is not constant along a vessel with elastic walls. The flow reduces downstream during vessel expansion and increases during contraction. This mechanism is used by compliant vessels to smooth out the sharp flow accelerations, accumulate the blood volume when the vessels expand and release volume during contraction. The law of continuity (1.2) states that the difference in the flow-rate between the two consecutive cross-sections must balance the change of the internal volume in that segment of the vessel. For example, the volume stored laterally in a segment with length L of an expanding vessel is $\Delta V=\Delta A\times L$, where ΔA is the change in the vessel area. Continuity (Eq. 1.2) also states that the rate of change of volume dV/dt, which is equal to $dA/dt\times L$ balances the difference of the flow rate, ΔQ, between the two ends of the vessel segment. As a result, the (negative) flow gradient along the vessel, $\Delta Q/L$, is balanced by the rate of expansion of the cross-sectional area of vessel, dA/dt (Fig. 1.2).

In general, by considering an arbitrarily brief segment, the flow gradient $\Delta Q/L$ along the vessel is expressed by dQ/dx, and the law of continuity for the vessel describes as

$$\frac{\partial Q}{\partial x}+\frac{\partial A}{\partial t}=0 \tag{1.3}$$

Equation 1.3 shows how a sharp peak of flow decreases downstream ($dQ/dx<0$) along an elastic (compliant) vessel due to vessel expansion ($dA/dt>0$) based on a synchronous peak of pressure (Pedley 1980, Sect. 2.1.1; Fung 1997, Sect. 3.8).

The law of continuity has been expressed in Eq. 1.2 in terms of a balance for the entire fluid volume, while Eq. 1.3 describes a balance across a cross-section of a duct. The same law holds for any arbitrary small volume of a moving fluid. Intuitively, conservation of mass implies that the fluid cannot move away from a given point along all directions unless additional volume of fluid replaces for it. In other words, the flow cannot *diverge* from a point. As an example, consider a microscopic volume as sketched in Fig. 1.3, continuity requires that the flow diverging from such volume along one direction will be balanced by the flow converging to

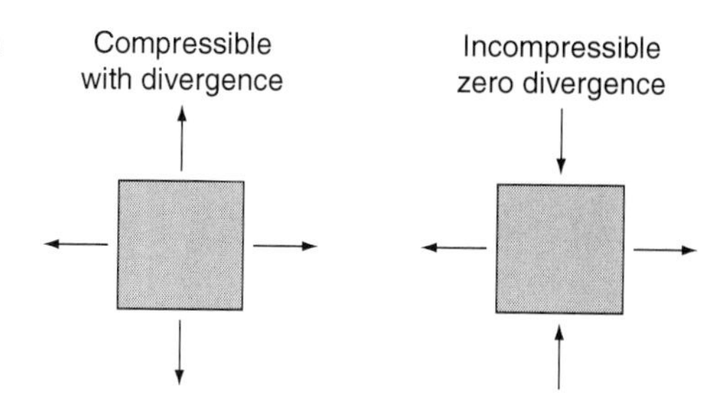

Fig. 1.3 The divergence of flow creates a deficit of mass inside the generic volume that can be accommodated only by a change of the flow density. In an incompressible fluid, the flow density cannot vary, and the contributions to mass along the different directions must balance and result in a total zero divergence of the flow

the same volume from another direction. In general, at any point in a flow field, the total *divergence* of the fluid motion must be equal to zero.

This concept introduces an important constraint on the physically realizable patterns of flow motion. Such a constraint facilitates the development of coherent vortices in the flow. When the velocity is constant it automatically satisfies the continuity since divergence can only occur in the presence of spatial velocity gradients (see Panton 2005, Sect. 5.1). In Fig. 1.3, the small volume presents a mass deficit along the direction x due to an increase in the velocity along that direction. This is a positive velocity gradient $\partial u_x/\partial x$. When the gradient is zero, the velocity at the two faces is equal and the same flow that enters one side exists from the other. Such a gradient measures the total flow-rate per unit volume, entering across the two surfaces facing the direction x. Similarly, the gradient, $\partial u_y/\partial y$, is the total flow-rate across the surfaces facing the direction y. In an incompressible fluid, the total flow-rate contributing to a fixed volume must be equal to zero. Positive gradient must be balanced by opposite-sign gradient along the other directions. In Equation

$$\frac{\partial u_x}{\partial x} + \frac{\partial u_y}{\partial y} + \frac{\partial u_z}{\partial z} = 0 \tag{1.4}$$

the velocity vector field has zero divergence. Equation 1.4 is often written in a more general and compact form as

$$\nabla \cdot u = 0 \tag{1.5}$$

using the pseudo-vector operator *nabla*.[1]

The zero-divergence characteristic is very important for the velocity field in cardiovascular system as it drastically reduces the adverse behavior of blood trajectories. Due to this property, vortices can be only developed along the regions where the flow is in contact with a wall.

[1] The generalization of equations in three dimensions often makes use of the vector-operator ∇, called *nabla*, for simpler compact writing. Nabla is a three-component derivative operator that can be seen as a vector of derivatives whose components in Cartesian coordinates, are given by $[\partial/\partial x, \partial/\partial y, \partial/\partial z]$.

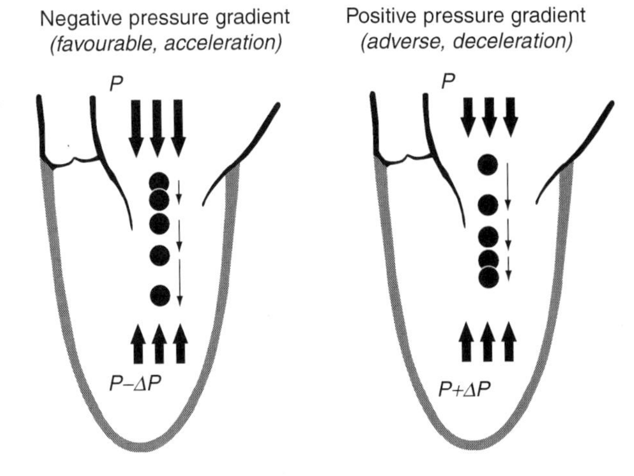

Fig. 1.4 A fluid particle accelerates in presence of a negative pressure gradient (*left picture*) and decelerates in presence of an adverse pressure gradient

1.3 Conservation of Momentum and Bernoulli Theorem

Momentum is a vector quantity equivalent to the product of the mass and the velocity. The conservation of momentum states that the impulse of any system does not change unless an external force is applied to it. The conservation of momentum can be expressed for a fluid by considering that a flow particle accelerates only when there is a pressure gradient applied to it. As sketched in Fig. 1.4, a fluid particle accelerates along a streamline when there is a *negative pressure gradient* along such a direction. Newton's second law per unit volume of fluid is shown here:

$$\rho a = -\frac{\partial p}{\partial s} \tag{1.6}$$

where ρ is the fluid density. However, the acceleration of a fluid particle, a, is not a realistically measurable quantity since individual particles cannot be followed during their motion. Therefore, it is only feasible to deal with quantities measured at fixed positions, rather than moving particles. As a result, the acceleration of a fluid particle is expressed in terms of velocity space-time variations at fixed positions.

Let us consider a particle that instantaneously passes through a fixed position x at the time t. With reference to Fig. 1.5 (left), consider a flow field with velocity that is spatially uniform and increases only in time. The fluid particles in this field accelerate while they cross the position x as the velocity increases everywhere. This acceleration given by the local velocity time-derivative $\partial u/\partial t$ is called *inertial acceleration* since it is associated with a change in the inertia of a volume of fluid. A particle may also accelerate in a steady flow when the velocity is constant everywhere in time, if it moves from a region of low-velocity toward a region where velocity is higher. The Fig. 1.5 (right) illustrates a particle that accelerates when

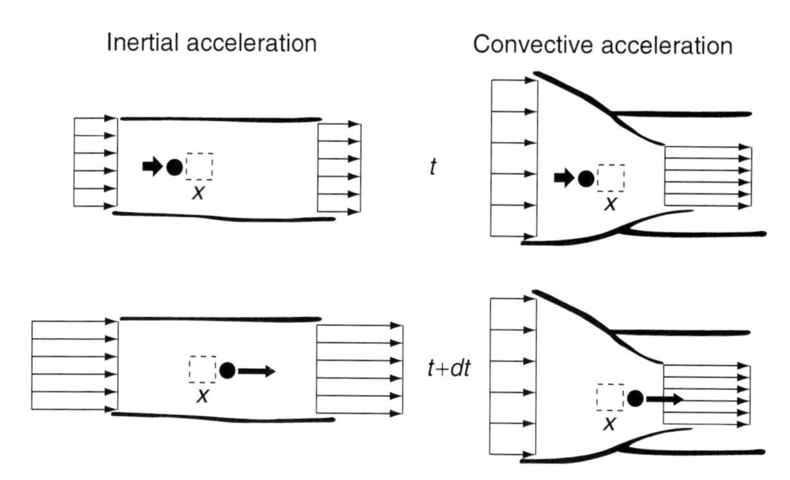

Fig. 1.5 Acceleration of a fluid particle measured in terms of velocity variations around the fixed position x. A particle accelerates when its velocity increases while crossing the position x. This can occur for inertial acceleration (*left panel*) when velocity at all places is increasing during time. Acceleration can also occur due to convection (*right panel*) when a particle moves toward a region where velocity is higher

enters into a section with smaller diameter. In other words, a particle accelerates if the velocity increases along the direction of the particle motion. This is called *convective acceleration*, which considers a particle that crosses the position x traveling a small distance $ds = udt$ during the time interval dt. If the velocity increases along the length ds, the particle experiences an acceleration based on the rate of change of velocity ($du/dt=udu/ds$).

In general, acceleration occurs based on two possibilities: (a) inertial acceleration, a reflecting the local time increase of velocity, and (b) convective acceleration, reflecting an increase in velocity in the direction of motion. In mathematical terms, the Newton law of motion (1.6) zfor the fluids translates as:

$$\rho\frac{\partial u}{\partial t} + \rho u\frac{\partial u}{\partial s} = -\frac{\partial p}{\partial s} \tag{1.7}$$

that is known as the *Euler equation*. This equation, although physically represents the balance of momentum for a moving particle, contains gradients of velocity and pressure, measurable at fixed points only. This equation requires some underlying assumptions; the first is that s is the direction of motion along a streamline, the second is that there is no frictional forces, and lastly no external force is being applied to the fluid.[2]

[2] This equation is also valid in presence of gravity (or any other *conservative* force).

The Bernoulli law is derived from the Euler equation. The Euler Eq. 1.7 can be rewritten in the alternate form of

$$\rho\frac{\partial u}{\partial t}+\frac{\partial}{\partial s}\left(\frac{1}{2}\rho u^{2}+p\right)=0 \tag{1.8}$$

which shows that the variation in the total energy along a streamline is balanced by the change of fluid inertia. Considering the integration rule for the sum of the variations along a line, integration of Eq. 1.8 between two arbitrary points of 1 and 2 along a streamline gives:

$$\frac{1}{2}\rho u_2^2+p_2=\frac{1}{2}\rho u_1^2+p_1-\rho\int_1^2\frac{\partial u}{\partial t}dx \tag{1.9}$$

which is called the general *Bernoulli equation*. The Bernouli equation represents the conservation of mechanical energy given by the sum of the potential energy, p, and the kinetic energy, $½\rho u^2$. The Bernoulli theorem states that variation of the mechanical energy from a point to another point along a streamline results in acceleration or deceleration of the fluid along that path, if there is no frictional (viscous) forces. The last term accounts for variation in the fluid inertia.

When inertial effect is negligible, the Bernoulli theorem reduces to a special case of the conservation of energy, $p+½\rho u^2$, along a streamline. The Bernoulli theorem is often employed to measure the pressure drop from a large cardiac chamber (when u_1 is negligible), at peak systole or at peak diastole when the inertial term is negligible. The same principle can be used for flow across the aortic valve or across a stenosis. In these cases, Eq. (1.9) simplifies to

$$p_2-p_1=\frac{1}{2}\rho u_2^2 \tag{1.10}$$

This formula is often utilized to obtain pressure from flow velocity, and if the velocity is expressed in terms of *m/s* and pressure in terms of *mmHg*, then $p_2-p_1=4u^2$.

The Bernoulli theorem represents the conservation of mechanical energy and allows evaluating the transformation of potential energy into kinetic energy and vice versa. It explains the underlying connection between variations in velocity and pressure. In its simplest form, flow that enters a stenotic segment increases its velocity due to conservation of mass and, because of the conservation of mechanical energy expressed by the Bernoulli theorem, reduces the pressure to balance the increase in kinetic energy.

In fact, the law of conservation of momentum implies that of conservation of energy and vice versa. In mechanics, momentum and energy balances are equivalent physical laws. The Euler equation (1.7) is an important simplified form, which is valid along a streamline only. For completeness, let us describe the three-dimensional Euler equation as:

$$\rho\frac{\partial u}{\partial t} + \rho u\cdot\nabla u = -\nabla p \tag{1.11}$$

written in compact form using the operator ∇, where the velocity, u, is a three-dimensional vector. The Euler equation (1.11) is either a vector equation or a system of three scalar equations along each coordinate. The Euler equation describes the motion of a fluid under the fundamental assumption that any form of friction is neglected, thus it represents the conservation of mechanical energy, allowing transformations between different states with no energy loss. This equation is valid for ideal fluids or *inviscid* flows in which the friction due to viscous phenomena can be neglected. These dissipative phenomena are considered in the next sections.

1.4 Conservation of Momentum and Viscosity

Viscosity is an intrinsic property of a fluid that gives rise to the development of viscous shear stresses inside the flow. Energy dissipation due to viscous stresses is the only mechanism for energy loss in fluids. In fact, the viscous friction phenomenon transforms mechanical energy into thermal energy.

The viscous stresses are responsible for the resistances of the fluid against motion. These stresses develop in presence of a velocity difference among the adjacent fluid elements that slide on each other. One example on the development of viscous friction is the confined fluid between two parallel plates, as shown in Fig. 1.6, where one plate moves relatively to the other. This motion is steadily sustained once a constant force is applied to the upper plate. This force is not associated with the acceleration of fluid particles and is balanced by the viscous friction. In such an arrangement, a fluid element is subjected to a forward traction exerted by the faster moving fluid above, and a backward resistance from the slower fluid below. To achieve equilibrium, the two stresses must be equal and opposite, given that the element does not accelerate and no other forces is applied. Such a flow presents a constant shear stress across a fluid gap. This value is equal to the wall shear stress acting on the moving plate and corresponds to the force per unit surface applied to keep it in steady motion.

In a Newtonian fluid, this shear stress is proportional to the shear rate that is equal to the ratio of the wall velocity U to the height of the fluid gap. The proportionality constant is the *dynamic viscosity*, commonly indicated with the symbol μ, whose normal value in blood is about 3.5×10^{-3} kg m^{-1} s^{-1}. The shear stress τ on a slice of a moving fluid is proportional to the shear rate across the slice surface:

$$\tau = \mu\frac{du}{dn} \tag{1.12}$$

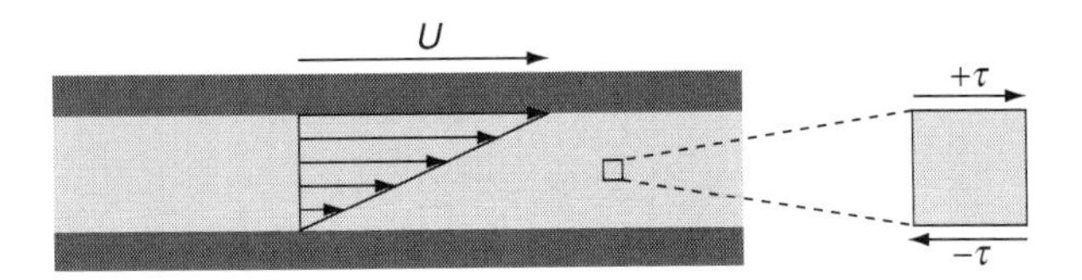

Fig. 1.6 When the upper plate is in motion with velocity U relative to the fixed plate below, shear stress is constant across the gap between the two plates because every fluid element is in equilibrium subjected to the shear forces given by the faster fluid above and the slower fluid below. In a Newtonian fluid, the velocity grows linearly from zero adjacent to the bottom plate to U at the upper plate

In a non-Newtonian fluid, the viscosity is not constant and depends on the local state of shear rate of the fluid elements. However, as previously discussed in Sect. 1.1, non-Newtonian effects are typically negligible in large vessels due to the dominance of the inertia in that setting. As a result, the non-Newtonian corrections do not significantly affect the analysis involved in vortex dynamics.

Newtonian viscous stresses can be immediately accounted for the motion along a streamline. When shear stress is constant across a streamline, a forward shear stress acts on that streamline from above and an identical backward stress acts from below with no net force (Fig. 1.6). In general (e.g. Fig. 1.4), the motion of the fluid particle is subjected to a shear stress on one side and another shear stress on the opposite side with a negative sign. The total viscous force, per unit volume, is thus due to the variation of the shear stress across the streamline, in formulas ($\partial\tau/\partial n$), where n indicates the direction across that streamline. This viscous force can be described in terms of the velocity, $\mu\partial^2u/\partial n^2$, based on Eq. 1.12. Considering the viscous forces in the equation of motion, it transforms the Euler equation into the Navier-Stokes equation:

$$\rho\frac{\partial u}{\partial t}+\rho u\frac{\partial u}{\partial s}=-\frac{\partial p}{\partial s}+\mu\frac{\partial^2 u}{\partial n^2} \tag{1.13}$$

The two equations differ by the viscous term only.[3] This term is not easy to evaluate in most flow conditions, and often neglected when applying the Bernoulli balance. However, in three-dimensional flow, the viscous term must be properly accounted for all gradients of shear stress across a point. As a result, the three-dimensional Navier-Stokes equation describes as (see Panton 2005, Sects. 6.2 and 6.6):

$$\frac{\partial u}{\partial t}+u\cdot\nabla u=-\frac{1}{\rho}\nabla p+\nu\nabla^2 u \tag{1.14}$$

[3] This intuitive result is somehow simplified: the viscous term should include variations along all directions about the streamline. In particular for three-dimensional flow, it should include the other direction perpendicular to both the streamline and to n. It was simplified here to avoid unnecessary symbolic complications.

This fundamental equation describes all aspects of the fluid motion, under the assumption of incompressible, Newtonian fluid, in the absence of non-conservative forces. The *kinematic viscosity* ν, which is introduced in Eq. 1.14 is given by the ratio of the dynamic viscosity to the fluid density, $\nu = \mu/\rho$, and its value for blood is about 3.3×10^{-6} m^2/s (or 3.3 mm^2/s). The kinematic viscosity is a more common viscous coefficient when dealing with blood motion, including vortex dynamics, because incompressible motion is independent of the actual value of density. This is a value that enters into play only as a multiplicative factor when motion is quantified in terms of force, work, or energy.

The kinematic viscosity is a small coefficient by itself. Therefore, for the viscous term to show significant effect, presence of sharp velocity gradients are required. Indeed, viscous effects are often negligible with respect to potential-kinetic energy transformation along brief paths (e.g. when the Bernoulli equation is used). Alternatively, dissipation is going to have significant effect once summed up over long fluid paths and for circulation balances.

Presence of viscosity introduces a fundamental novel element to fluid dynamics. Viscosity implies the continuity of motion between adjacent slices that smoothly slide over each other due to the gradient of velocity. However, they cannot present a net velocity difference otherwise the shear rate and the viscous term would rise to infinity. This continuity also applies to the first fluid elements adjacent to a solid boundary, where it implies the *adherence* between the fluid and structure. This condition, normally referred as *no-slip condition*, is a result of viscosity and does not apply in an ideal fluid.

1.5 Boundary Layer and Wall Shear Stress

The adherence of fluid at the solid boundaries is a purely viscous phenomenon, which implies that the viscous effects can never be neglected close to the walls. Near the solid boundary, a layer of fluid exists whose motion is directly influenced by the adherence to the boundary. This influence of wall on the fluid is gradually reduced, and sometimes even disappears once moving away from the boundary toward the bulk flow. The *boundary layer* is the region where velocity rapidly grows from a zero value at the wall to reach values comparable to those found at the center of the vessel. The *wall shear stress*, τ_w, is the stress exerted by the fluid over the endothelial layer, and is proportional to the wall shear rate, which is the gradient of velocity at the wall. This value is approximately the ratio of the fluid velocity away from the wall, U, to the boundary layer thickness, commonly indicated as δ:

$$\tau_W = \mu \frac{du}{dn}\bigg|_{wall} \approx \mu \frac{U}{\delta} \tag{1.15}$$

The higher the fluid velocity is, the higher the wall shear stress will be. Additionally, a thinner boundary layer results in higher wall shear stress.

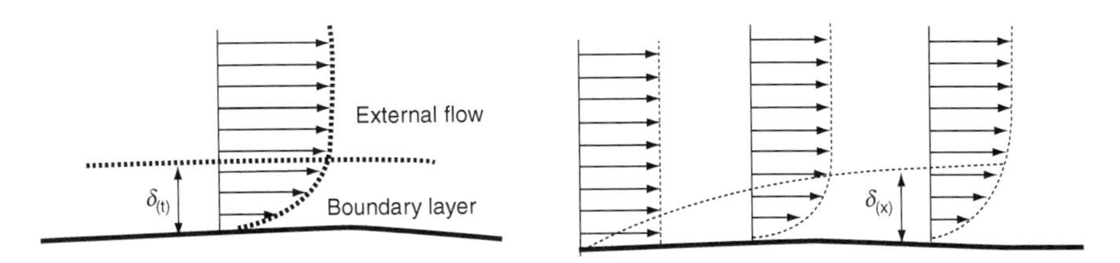

Fig. 1.7 The boundary layer is the thin region between the wall and the external flow. It grows in time in a starting flow (*left*), and it grows downstream of the entrance of a vessel (*right*)

The thickness of the boundary layer, δ, grows based on the tendency of shear to diffuse away from the wall due to the viscous friction between adjacent fluid elements. Let us consider a fluid that is initially at rest being impulsively set to motion; at the very beginning, the whole fluid volume is set into motion with the exception of the first layer next to the wall where the fluid particles remain adherent. As time proceeds, this viscous adherence slows down the adjacent layers and the boundary layer progressively grows, as shown in Fig. 1.7 (left). Its rate of growth is directly proportional to the kinematic viscosity ν, and inversely proportional to the thickness of the layer ($d\delta/dt \approx \nu/\delta$), which results in a square root growth in time.

$$\delta(t) \approx 5\sqrt{\nu t} \tag{1.16}$$

The coefficient in front (here set equal to 5) may vary in different types of flows. Eq. 1.16 represents a general expression for estimating the length reached by a viscous diffusion process.

Based on the same reasoning, the boundary layer growth downstream the entrance of a vessel can be evaluated; as shown in Fig. 1.7. (Schlichting and Gersten 2000, Sect. 2.2; Fung 1997, Sect. 3.5)

$$\delta \approx 5\sqrt{\nu \frac{x}{U}} \tag{1.17}$$

The size of the boundary layer cannot grow indefinitely as discussed earlier based on the Eqs. 1.16 and 1.17. For example, in a vessel, the boundary layer can grow until it fills the entire vessel. Once its thickness is comparable to the vessel radius, it has no room for further increase. This is the particular case of the *Poiseuille flow* (see Sect. 1.6). Otherwise, for oscillatory cardiac flow, the boundary layer develops during one heartbeat and restarts from zero in the next one. Therefore, its size growth is limited to a fraction of the heartbeat.

Boundary layers are important because they are the locations where the shear stress develops due to the frictional forces on the surrounding wall. These boundary regions are often unstable and can detach from the wall, penetrating the bulk flow regions with high velocity gradients. In fact, the *boundary layer separation* is the only mechanism that generates vortices in incompressible flows.

1.6 Simple Flows and Concepts of Cardiovascular Interest

A simple but important type of flow motion is a flow inside a cylindrical vessel with circular cross-section. Considering that the flow is steady and uniform, both inertial and convective accelerations are zero (see Eq. 1.13). To be more precise, a cylinder of fluid with unit length and radius r is pushed ahead by the negative pressure gradient dp/dx acting on the cross-sectional area πr^2, and is subjected to the shear stress $\tau = \mu du/dr$ on the lateral surface $2\pi r$. These two forces must balance to ensure equilibrium. Therefore:

$$\frac{du}{dr} = \frac{r}{2\mu}\frac{dp}{dx} \tag{1.18}$$

where the pressure gradient is constant over the cross-section because no cross-flow exists. The Eq. 1.18 is satisfied by a parabolic velocity profile $u(r)$ that ensures the adherence condition at the vessel wall (Fig. 1.8). Assuming a vessel with radius R, the solution is

$$u(r) = \frac{-1}{4\mu}\frac{dp}{dx}\left(R^2 - r^2\right) = \frac{2U}{R^2}\left(R^2 - r^2\right) \tag{1.19}$$

Equation 1.19 is the well-known *Poiseuille flow* (see Schlichting and Gersten 2000, Sect. 5.2.1; Fung 1997, Sect. 3.2). The *Poiseuille flow* can be equivalently expressed in terms of the pressure gradient or the mean velocity U (second equality in Eq. 1.19), that are related based on $dp/dx = -8\mu U/R^2$ verifiable by taking an integral from Eq. 1.19. It features a maximum velocity at the center that is *twice* the average velocity, and a wall shear stress of $4\mu U/R$. Pressure loss, $-dp/dx$, is usually expressed with respect to the kinetic energy, $\frac{1}{2}\rho U^2$, per unit diameter length, $D = 2R$ of the tube. The *friction factor*, λ, is described as:

$$\lambda = \frac{-2D}{\rho U^2}\frac{dp}{dx} \tag{1.20}$$

that is a dimensionless quantity suitable for generalization under different conditions. In the Poiseuille flow, the friction factor is given by $64/Re$ where the *Reynolds number* is

$$Re = \frac{UD}{\nu} \tag{1.21}$$

and represents the ratio of inertial to viscous forces in the fluid. Flows with low values of the Reynolds numbers are highly viscous, smooth, with high pressure-

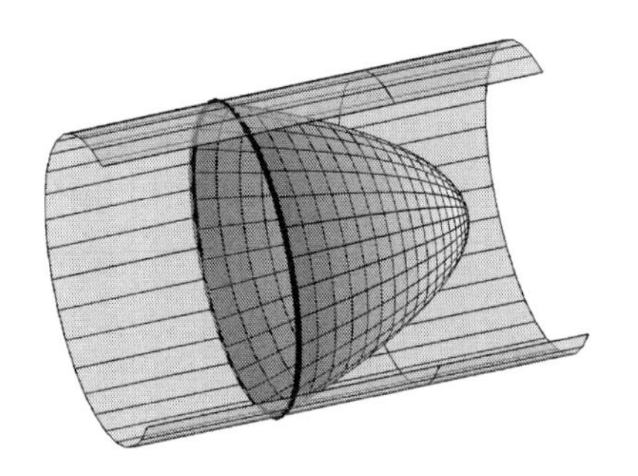

Fig. 1.8 The parabolic velocity profile or Poiseuille flow, develops in a cylindrical duct with circular fixed walls. It is valid under the assumption of steady flow and represents a good approximation of the time-averaged flow

loss. Flows at higher *Re* exhibit less dissipation to the available kinetic energy, thus they are less smooth and more easily subject to instabilities.

The Reynolds number has a fundamental role in the stability of any fluid motion. Once its value increases above a certain critical threshold, the flow does not dissipate the incoming energy and is prone to instability toward a more complicated motion to achieve further dissipation. The critical value for Poiseuille flow is about $Re_{cr}=2{,}300$. Above this value, the rectilinear flow becomes unstable. In that case, the flow is considered in transition to turbulence. However, fully turbulent motion is anticipated at *Re above* 10000. Normally, in the cardiovascular system, the Reynolds number is transitional only at mid-diastole in the LV and mid-systole in the aorta. The peak *Re* may even reach about 7000.

Cardiovascular flow is not steady but pulsatile, and the Poiseuille profile (1.19) is valid only where pulsatility is negligible. When studying uniform, unsteady flow, the momentum balance (1.18) must include the inertial acceleration due to the velocity variation in time. A featuring result of the pulsatile flow is that the unsteady boundary layer has a thickness (following Eq. 1.16) proportional to $\sqrt{\nu T}$ where T is the duration of the pulse. When the oscillation is rapid, the inertial term is not negligible, and a thin boundary layer develops near the wall that gives rise to a solution that fundamentally differs from the parabolic profile seen in Poiseuille flow.

The *Womersley number* is the ratio of the vessel diameter, D, to the thickness of the unsteady boundary layer (see Fung 1997, Sect. 3.5):

$$Wo = \frac{D}{\sqrt{\nu T}} \tag{1.22}$$

This number is a useful parameter to discern the behavior of unsteady flows. When *Wo* is close to or smaller than 1, the viscous friction diffuses from the wall into the entire vessel. The unsteady flow is a sequence of parabolic profiles of Poiseuille type, as shown in Fig. 1.9. Given that the denominator of (1.22) cannot significantly vary in humans, with values little below $2\,mm$, the parabolic solution is found in all vessels whose diameter does not exceed this value. In larger arteries, *Wo* can be around or even above 10, for example in the aortic root. In these cases, as shown in Fig. 1.9, the oscillatory boundary layer is confined to a fraction of the vessel radius, and produces relatively sharp local variation in the velocity profile adjacent to the

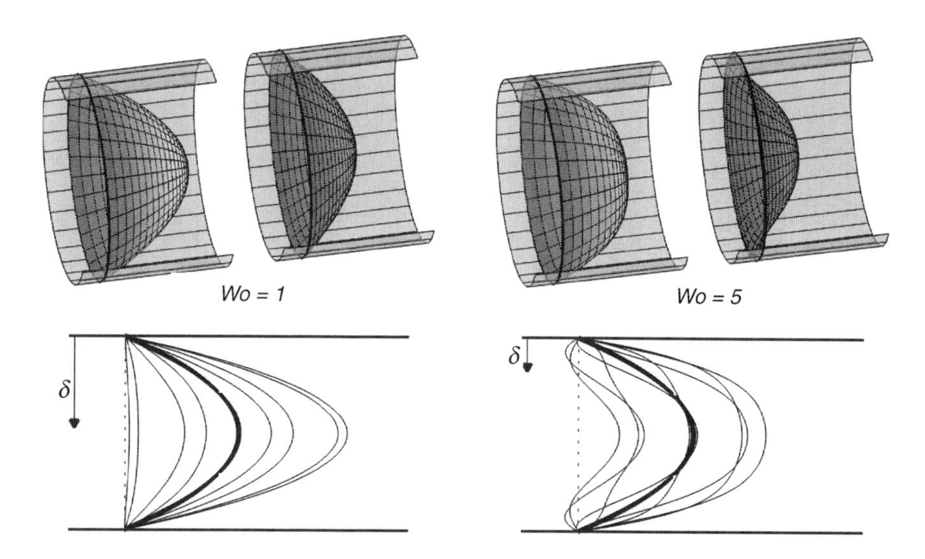

Fig. 1.9 Examples of the unsteady velocity profile that develops in a cylindrical duct when the flow is pulsatile. The *left* side corresponds to a Womersley number $Wo = 1$, which means that the boundary layer has a thickness comparable to the vessel radius. The *right* side corresponds to $Wo = 5$, in which the oscillatory boundary layer is smaller and remains localized close to the wall. The spatial velocity profile is shown on top during peak flow deceleration. The flow profile at eight instants during one heartbeat are shown below, the thick line corresponds to the time-average Poiseuille profile

wall. The Poiseuille parabolic profile is a good approximation for the time-averaged flow characteristics only, while calculation of extreme values requires the evaluation of the oscillatory terms.

An analogous modification of the flow due to the presence of a boundary layer with limited thickness is found at the proximal region of vessels. Here, the parabolic, steady, or time-averaged solution does not establish. At the entrance of a vessel, the boundary layer starts from zero and grows downstream "the entry region" where the boundary layer has not yet grown enough to fill the entire vessel (Fig. 1.7). The velocity profile is flatter in the bulk region, and not reached by the boundary layer while the centerline velocity is initially equal to the average velocity U and slowly increases toward the Poiseuille value, $2U$. The length of this "entry region" can be obtained from Eq. 1.17.

The previous simple examples of flow in straight vessels provide an insight on how the shear that develops in the boundary layer may influence the overall flow, and its interaction with the wall. The effect of the boundary layer is even more significant in presence of irregularly-shaped geometries. A simple example is the flow that crosses a brief constriction such as a stenosis or an orifice. At the constriction, the boundary layer does not develop other than along a very short length with a tiny thickness. However, the boundary layer exists to ensure the adherence condition, although hardly visualized. Such a small, intense boundary layer takes a fundamental role in the subsequent flow development.

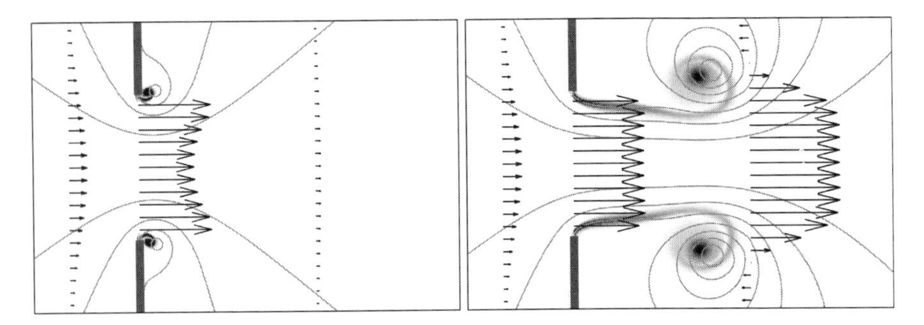

Fig. 1.10 Two snapshots during the development of flow crossing a circular orifice. At the initial stage, *left panel*, the fluid velocity is higher near the edges because of the convergence toward the constriction. The boundary layer at the orifice is extremely thin because it has no development wall upstream. As time proceeds, the jet extends downstream the core region bounded by a shear layer. The pictures show the velocity profile at three positions (*same scales*), the streamlines (*gray lines*), and the shear region (*shaded gray*) developing due to the boundary layer

To further clarify this concept, let us consider the flow for an extreme case of the irregular vessel. At the initial instant of a starting flow across an orifice, as depicted in the left panel of Fig. 1.10, the flow converges from the chamber, upstream into the orifice and then diverges in an approximately symmetrical manner. Nevertheless, the tiny boundary layer is convected downstream the orifice with the moving fluid. Initially, such a shear layer travels a very short distance (Fig. 1.10, left side), and its tip rolls-up because of the velocity difference between the two sides. As time proceeds, the original boundary layer gives rise to an elongated free *shear layer* that surrounds the jet core and separates it from the ambient fluid (Fig. 1.10, right side). The forefront jet is composed of the initially separated shear layer that has rolled-up into the jet front.

The phenomenon described here represents a typical mechanism for formation of vortices. It is a consequence of viscous friction that develops from the adherence at the wall, as discussed earlier in Sects. 1.4 and 1.5. The viscous effects are confined to the shear layer only, and the other flow regions that are not directly influenced by the shear layer elicit negligible viscous effects. This fact describes how very small regions with dominant shear phenomena may dramatically change the overall flow field. This behavior is commonly observed, and is evident in vortical flow patterns where vortices cover limited regions in the space at particular times.

References

Fung YC. Biomechanics: circulation. New York: Springer; 1997.

Panton RL. Incompressible flow. 3rd ed. Hoboken: Wiley; 2005.

Pedley TJ. The fluid mechanics of large blood vessels. Cambridge: Cambridge University Press; 1980.

Schlichting H, Gersten K. Boundary layer theory. 8th ed. Berlin Heidelberg: Springer; 2000.

Chapter 2
Vortex Dynamics

Abstract This chapter introduces the fluid motion in terms of vortex dynamics. First, the conceptual background on vortex dynamics is presented to attain an intuition why most fluid phenomena involve vortices and why they are so relevant. Then the origin of vortices, their interaction and dissipation will be discussed. Vortices reveal to have major effects on the wall shear stress along the nearby structures, and the process of vortex formation is associated with the development of forces on surrounding boundaries.

2.1 Definitions

Vortices are fundamental performers in fluid mechanics and they develop in almost every realization of fluid motion. The presence of vortices dominates the current understanding from fluid dynamics and the associated energetic phenomena. Vortices that develop in the large vessels of the cardiovascular system play a fundamental role in the normal physiology and bring about the proper balance between blood motion and stresses on the surrounding tissues.

The fluid velocity is commonly assumed as the principal quantity describing fluid motion. However, velocity is not able to evidence the underlying dynamical structure of a flow field such as stresses, mixing, or turbulence that depend on velocity gradients. The weakness of a description solely based on velocity is particularly revealed when the fluid motion exhibits the presence of vortex structures. In general, *vorticity* is another fundamental quantity used for the analysis of incompressible fluid dynamics. Vorticity, which represents the local rotation rate of fluid particles, allows emphasizing on the flow structures that are in the background of the flow field. It also represents a complete description of the flow and allows recovering the whole velocity field.

A. Kheradvar, G. Pedrizzetti, *Vortex Formation in the Cardiovascular System*,
DOI 10.1007/978-1-4471-2288-3_2,

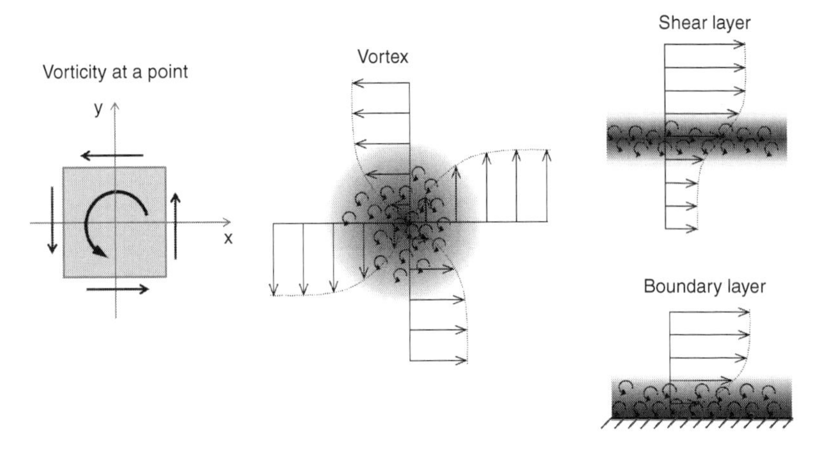

Fig. 2.1 Vorticity corresponds to the local rotation of a fluid particle. The spatial distribution of vorticity gives rise to different flow structures. An accumulation of vorticity in a compact region corresponds to a vortex; an elongated distribution of vorticity corresponds to a shear layer that, when it is adjacent to the wall, is a boundary layer

In mathematical terms, the vorticity is a vector, indicated with $\omega(t,x)$ and defined as the *curl* of the velocity field $u(t,x)$, which is normally expressed as the cross product of the *nabla* operator with the velocity field

$$\omega(t,x) = \nabla \times u \tag{2.1}$$

that in Cartesian coordinates reads as:

$$\begin{bmatrix} \omega_x \\ \omega_x \\ \omega_x \end{bmatrix} = \begin{bmatrix} \dfrac{\partial u_z}{\partial y} - \dfrac{\partial u_y}{\partial z} \\ \dfrac{\partial u_x}{\partial z} - \dfrac{\partial u_z}{\partial x} \\ \dfrac{\partial u_y}{\partial x} - \dfrac{\partial u_x}{\partial y} \end{bmatrix} \tag{2.2}$$

The interpretation of vorticity is particularly intuitive in a two-dimensional flow field, when only the x and y components of the velocity field exist. In this case, vorticity has only the component z, perpendicular to the plane of motion, $\omega = \partial u_y/\partial x - \partial u_x/\partial y$, which physically corresponds to (twice) the local angular velocity of a fluid particle. In fact, a positive vorticity corresponds to a vertical velocity, u_y, increasing horizontally, along x, and a horizontal velocity u_x, decreasing vertically. This type of velocity differences about a point represents a rotational motion Fig. 2.1 (left) (Panton 2005, Chap. 3).

The significance of vorticity is not limited to local rotation. Vorticity is commonly considered the skeleton of the flow field and a fundamental quantity to define the flow structure. A *vortex* can be loosely described as a fluid structure that possesses circular or swirling motion; although a more proper definition is that of a compact region of vorticity. In addition to vortices, the vorticity map allows recognition of any basic flow structure. A shear layer is actually a layer of vorticity, or a *vortex layer*. The boundary layer discussed in the previous chapter is a vortex layer adjacent to the wall that develops due to the velocity difference between the outer flow and the fluid attached to the wall. The relationship between velocity and vorticity distributions are sketched in Fig. 2.1. The intensity of a vortex is normally measured by its *circulation*, commonly indicated by Γ. This quantity is the integral of the velocity along a closed circuit surrounding the vortex, and is equivalent to the sum of all the vorticity within an area of circulatory motion. The intensity of a vortex layer is measured by the difference in velocity, the *velocity jump* Δu, between the flow above and below the layer. Vortices and vortex-layers are the fundamental vorticity structure in flow fields. Their different three-dimensional arrangements give rise to the complex evolving flows.

The significance of vorticity can be best represented by the decomposition of the complete velocity field into two distinct contributions; a *rotational* component u_{rot} that accounts for the whole vorticity in the flow field, and a *irrotational* component u_{irr} that is independent from the vorticity content:

$$u = u_{rot} + u_{irr} \tag{2.3}$$

The irrotational component of the velocity field is a particularly simple field in incompressible flows. It is inferred from the conservation of mass only (continuity constraint), and does not involves the equation of motion. The irrotational flow helps satisfying the instantaneous balance of mass without any evolutionary mechanism through kinematic congruence only. Therefore, flow without vorticity gives rise to an irrotational velocity field and does not depend on the balance of momentum. When required, the equation of motion can be employed to derive the pressure distribution from the velocity. In the case of irrotational flow, this can be performed with the simple Bernoulli equation for an ideal flow because energy dissipation is absent in an irrotational flow. In fact, the viscous term of the Navier-Stokes (Eq. 1.14), $\nabla^2 u$, which can be expressed for an incompressible flow as $\nabla \times \omega$, is identically zero for a flow without vorticity.

These considerations on characteristics of the irrotational flow follow from the fact that the velocity can be expressed as the gradient of a scalar field, so-called potential φ, as $u_{irr} = \nabla \varphi$ (the curl of a gradient is identically zero). In an incompressible flow, the continuity Eq. 1.5 applied to such a gradient field becomes the well-known *Laplace equation* for the flow potential, $\nabla^2 \varphi = 0$. This is a linear equation whose solutions can be obtained based on several boundary value methods. Given the linearity of the Laplace equation, the irrotational velocity component can be expressed as a direct superposition of several elementary irrotational flows.

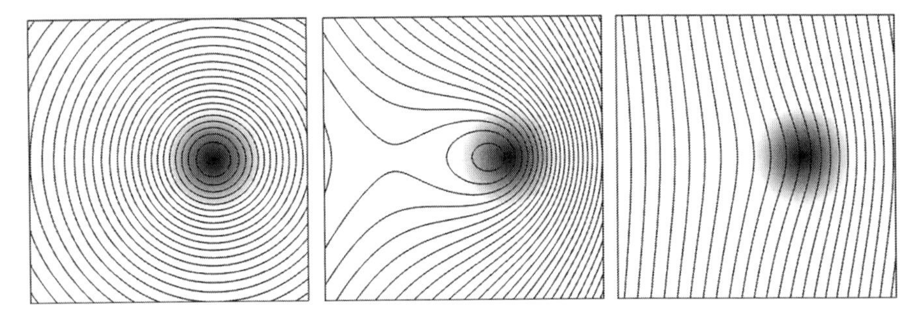

Fig. 2.2 A vortex is properly defined as a region where vorticity has accumulated. A definition based on the presence of circular streamlines may be ambiguous in the presence of a background flow. To further clarify, the three panels show the streamlines corresponding to a flow field with identical vorticity magnitude with additional uniform flow of increasing intensity. (*left*) A vortex alone gives rise to circular streamlines; (*central*) the streamlines are modified when a uniform flow with moderate intensity is added; (*right*) in presence of a uniform flow with high intensity, the streamlines do not show any rotational path. It should be mentioned that the actual vortex in the flow is identical in all the panels, and the velocity gradients and shear rate acting on the fluid are not modified as well

The velocity decomposition is the key approach to recognize the presence and the role of vortices in the flow. A vortex, as a region with accumulation of vorticity, is not necessarily a region exhibiting circulatory motion. The velocity field corresponding to an isolated vortex is purely rotational, and its streamlines rotate about the vortex core and describe a circulatory motion as shown in Fig. 2.2 (left). Once an irrotational contribution adds on top of the same vortex flow, it may modify the apparent vortex signature in terms of streamlines. To explain this, let us consider the same vortex of Fig. 2.2 (left) with the addition of uniform flow, a rigid translational motion from top to bottom that is evidently irrotational and does not affect the value of vorticity. The resulting flow fields are shown in Fig. 2.2 for increasing values of the uniform motion in the central and rightmost panels. The three fields in Fig. 2.2 present exactly the same vortex, with the same gradients of velocity at all points. However, the underlying vortex may not look similar, based on appearance of streamlines.

Fluid dynamics phenomena related to the rotational part of the velocity field (e.g., evolutionary dynamics, friction, dissipation, forces, boundary layer, vortex formation, etc.) usually dominates the irrotational contribution, which may affect the transport and mass conservation only.

2.2 Dynamic of Vorticity

The vorticity is a fundamental quantity that describes the tendency for fluid's elements to spin. The entire flow field inside a given geometry can be reconstructed based on knowledge of vorticity field only. From mathematics perspective, vorticity is a field with zero divergence since the divergence of a curl is zero:

$$\nabla \cdot \omega = 0 \tag{2.4}$$

This means that the vorticity field cannot take arbitrary geometric shapes. Vorticity typically develops in terms of vortex tubes or vortex layers. The total vorticity contained inside a vortex tube is conserved. A vortex tube cannot terminate abruptly, and must either be a closed ring or terminate by spreading into a vortex layer. The mathematical law governing the evolution of vorticity can be immediately derived from the conservation of momentum. The Navier-Stokes Eq. 1.14, can be rearranged in terms of vorticity; by taking the curl of the Navier-Stokes equation, the *vorticity equation* would be obtained (Panton 2005, Sect. 13.3):

$$\frac{\partial \omega}{\partial t} + u \cdot \nabla \omega = \omega \cdot \nabla u + \nu \nabla^2 \omega \tag{2.5}$$

This equation expresses the law of motion in terms of vorticity. Despite the apparent mathematical complexity, this equation allows extracting some important information regarding vortex dynamics. For example, it can be immediately recognized that the vorticity equation does not contain the pressure term or any conservative force like gravity. In fact, the distribution of pressure has no direct influence on vortex dynamics. On the contrary, pressure strongly depends on vorticity that results in friction and energy losses.

If vorticity is zero at an instant of time, it remains zero afterward. This is seen by inspection of Eq. 2.5 where all terms are identically zero once vorticity is zero, and cannot change in time. Given that the vorticity cannot be created inside the fluid, it can only be generated at the interface between the fluid and the boundary. This is a fundamental element for the study of vortex dynamics. In incompressible flows, vorticity does not appear spontaneously within the fluid.

Vorticity is subjected to several evolutionary phenomena. The primary one is that vorticity is transported with the flow as if it were a passive tracer. This is represented by the two terms on the left hand side of Eq. 2.5 that simply describe the variation of vorticity over a particle moving with the flow. The first term is the time-variation of vorticity at the fixed position crossed by the particle, and the second term describes how vorticity varies when the velocity is aligned with a positive gradient of vorticity. Therefore, vorticity moves with the local fluid velocity, like a tracer, and can further change its value in virtue of two additional phenomena.

The first term on the right hand side of Eq. 2.5 represents the phenomenon of increase in vorticity due to *vortex stretching*. Let us consider a small cylinder of fluid whose velocity increases along its axis. Thus the velocity is lower at the base and higher at the cylinder top. As time proceeds, the cylinder elongates, due to the velocity gradient and narrows in the transversal direction based on the conservation of mass. In a two-dimensional flow, the vorticity is perpendicular to the plane of motion with no velocity gradient out of plane; vorticity stretching is intrinsically a three-dimensional effect.

The last term in the right hand side of the Eq.2.5 contains the viscous effects. Before describing it, let us recapitulate the dynamics of vorticity in absence of viscosity. An element of fluid that contains no vorticity remains without vorticity afterward. This is the first of the three *Helmholtz's laws* for inviscid flow (Panton 2005, Sect. 13.9). The vorticity is a vector that behaves like a small-string element of fluid. It moves with the flow and is stretched and tilted with it. This is essentially the second Helmholtz's law. The third law indicates that the vorticity is a field with zero divergence (Eq. 2.4), and the total vorticity contained inside a vortex tube or a vortex filament is conserved along the filament while it moves with the flow.

In a two-dimensional flow, the vorticity vector has a unique non-zero component perpendicular to the plane of motion; stretching is absent and vorticity is simply transported with the flow. The magnitude of vorticity is stuck onto the individual fluid particles. Vorticity simply organizes into vortex patches, and redistributes into vortex layer, according to the motion of fluid particles.

The viscous term in the Eq. 2.5 introduces the effects of friction and energy dissipation in terms of vorticity. The effect of viscosity on vorticity is a phenomenon analogous to that of heat diffusion or diffusion of a tracer (Panton 2005, Sect. 13.4). The distribution of vorticity is smoothed out by viscosity as a sharp vortex progressively reduces its local strength while it widens its size in a way that the total vorticity is conserved. Like in any diffusive process, the rate of diffusion is higher in presence of sharp vorticity gradients. Therefore, the magnitude of viscous dissipation become increasingly relevant where vorticity presents changes over short distances. This leads to the most important aspect of energy loss in fluid motion, which is called viscous dissipation, and is most effective at small scales. The vector property of viscous diffusion is evidenced when it produces the annihilation of close patches of opposite sign vorticity. This has a peculiar consequence in three-dimensions. When two portions of vortex filaments get in contact, the opposite-sign vorticity locally annihilates. This accompanies the reconnection of the cropped, oppositely-pointing vortex lines that cannot terminate into the flow. The viscous reconnection phenomenon is the underlying mechanism leading to topological changes, metamorphoses of three-dimensional vortex structures, and increased dissipation because of turbulence (see Sects. 2.7 and 2.8).

2.3 Boundary Layer Separation

As discussed earlier, vorticity cannot be generated within the incompressible fluid. Vorticity can only develop from the wall due to the viscous adherence between the fluid and the bounding structure. Vorticity is produced because of the no-slip condition at the interface between the fluid and the solid surface. It then progressively diffuses away from the wall through the viscous diffusion mechanism to produce a layer of vorticity at the boundary. The boundary layer thickness corresponds to the length at which the viscous diffusion penetrates into the flow, is proportional to $\sqrt{\nu t}$ (Eqs. 1.16 and 1.17).

The boundary layer is fundamentally significant in fluid mechanics as it represents the *unique* source of vorticity in a flow field. It can be easily verified that the value of vorticity at the wall corresponds to the wall shear rate and is proportional to the wall shear stress

$$\tau_W = \mu\omega \tag{2.6}$$

The vorticity at the wall is often being referred as wall shear rate.

In small vessels, the thickness of the boundary layer is comparable to its diameter, and fills the entire flow field. At such small scales, in arterioles and venules, viscous diffusion is the dominant phenomenon. Therefore, vortices are absent. On the contrary, in large blood vessels or inside the cardiac chambers, the boundary layer often remains thin and is capable to penetrate for diffusion over a small fraction of the vessel size. Indeed, until it remains attached to the wall, it has a minor influence on the flow and only represents a viscous slipping cushion for the outside motion. However, under many circumstances, it happens that such a thin boundary layer detaches from the wall and is ejected into the bulk flow. This is the process of *boundary layer separation*, when thin layers with intense vorticity enter into the flow and give rise to local accumulation of vorticity and eventually to the formation of compact vortex structures.

Boundary layer separation is normally a consequence of the local deceleration of the flow (Panton 2005, Sect. 20.11; Batchelor 1967, Sect. 5.10). The whole process of boundary layer separation is sketched in Fig. 2.3. When flow decelerates, the boundary layer is subject to deceleration as well. Therefore, because of incompressibility, a local stream-wise deceleration associates with a growth of the thickness at the same location. This tongue of vorticity is lifted and strained by the outside flow once the vorticity value at the wall below it decreases. As this process progresses, opposite-sign wall-vorticity appears and a secondary boundary layer develops below the separating shear layer. The separation point at the wall, from where the separation streamline departs, corresponds to the place where vorticity is zero. The secondary vorticity is decelerated in its backward motion, and is lifted up. Eventually, it cuts the connection between the original boundary layer and the separating vorticity that detaches and enters into the flow. It should be also reminded that vorticity is not a passive tracer; it is made of velocity gradients and represents the underlying structure of the flow. Fig. 2.3 shows the velocity profiles and streamlines that develop due to the separating vorticity field.

Separation of boundary layer is thus a consequence of the local deceleration of the flow. In other terms, separation develops in presence of an *adverse pressure gradient* (i.e., a pressure growing downstream) that pushes from downstream and decelerates the stream. The most common way of having an adverse pressure gradient is due to a geometric change such as a positive curvature of the wall that occurs during an enlargement in a vessel. In this case, the velocity decreases for mass conservation, and the reduction of kinetic energy corresponds to a pressure increase due to the Bernoulli balance. Therefore, boundary layer separation develops behind a

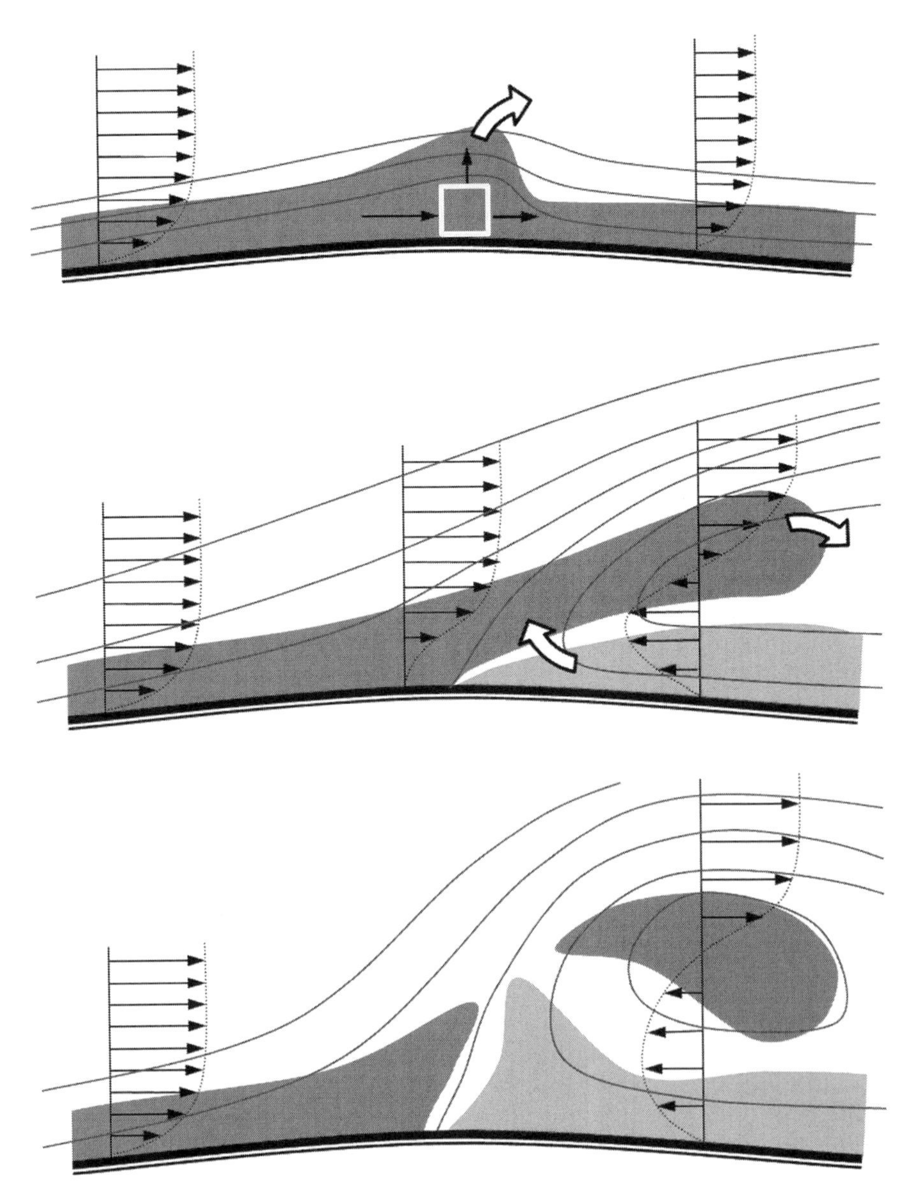

Fig. 2.3 The boundary layer separation process is schematically shown here. The dark gray color indicates layers with clockwise vorticity, the light gray color is the counter-clockwise; streamlines and velocity profiles are drawn. The flow deceleration produces a local thickening of the boundary layer due to mass conservation balance (*upper panel*). Such emerging vorticity is finally lifted and transported downstream by the external flow (see *arrows*). A shear layer then extends away from the wall and produces a secondary boundary layer, with oppositely-rotating vorticity (*mid panel*). The separated clockwise vorticity tends to roll-up while the secondary layer lifts up due to the same initial mechanism, as it backward motion is decelerating (see *arrows*). Eventually, the separating vortex layer detaches from the boundary layer and becomes an independent vortex structure

stenosis, or at the entrance of an aneurism. An extreme case of geometric change is that of a sharp edge, which is often found at the entrance of a side-branching vessel, and on the trailing edge of the cardiac valves. In the case of sharp edges, the flow deceleration is so local that the position of boundary layer separation is definitely localizable there. The vorticity that develops on the upstream side detaches at the sharp edge and leaves the structure tangentially.

Geometric changes are not the only possible sources for the development of flow deceleration. The velocity reduces together with an adverse pressure gradient that develops immediately downstream a branching point. Similarly, boundary layer separation develops due to the so-called *splash* effect, when a jet reaches a wall and produces high velocity streamlines that decelerate once they are deflected along the wall. Finally, the local flow deceleration is also produced by previously separated vortices. A vortex that gets close to a wall gives rise to a localized increase or reduction of the flow velocity at the wall below, and a corresponding deceleration immediately downstream or upstream, depending on its circulation. The vortex-induced boundary layer separation is a frequent phenomenon that may become particularly critical in some applications. In fact, the area of principal separation is often localizable and properly protected, whereas an unexpected separation due to a previously separated vortex may occur at unexpected locations.

2.4 Vortex Formation

The separation of the boundary layer represents the starting phase of the vortex formation process. The featuring characteristic of any shear layer is the velocity difference between its two sides. The farther side of a shear layer that detaches from the wall moves with a speed that is higher than the side closer to the wall. Therefore, the separating shear layer curves on itself and eventually rolls-up into a tight spiral shape. During the rolling-up process, the distance between the two successive turns of the vortex layer progressively reduces with the closest neighboring turns at the center of the spiral. The viscous diffusion process smears out the tight spiraling structure into a compact inner core with a smooth distribution of vorticity (Wu et al. 2006, Sect. 8.1).

The roll-up and formation of an isolated vortex behind a sharp-edge obstacle are shown in Fig. 2.4. In the case of such a sharp geometric change, the boundary layer separation localizes at the edge, and the boundary layer from the upstream "wetted" face of the obstacle leaves the edge tangentially, immediately rolls-up into a spiral. The vorticity viscous diffusion acts with higher strength in the tighter inner spiral branch and gives rise to a smooth vortex core.

An isolated forming vortex grows with a self-similar shape until there are no external disturbances that can influence its formation until the vortex size is small enough in comparison to the size of the surrounding geometry. The properties of the initial self-similar growth can be obtained by simple dimensional arguments.

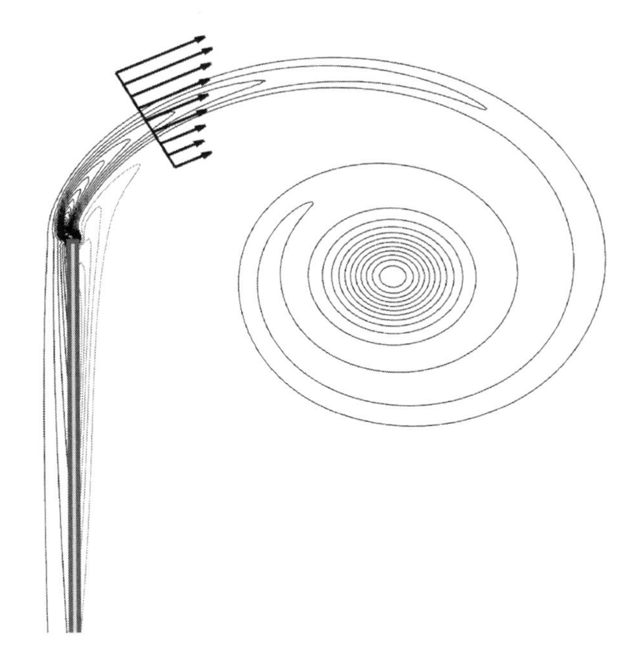

Fig. 2.4 Vortex formation from a sharp edge obstacle. The shear layer separates from the upstream "wetted" side of the wall and rolls-up into a spiral. The tight turns in the inner part of the spiral spread for viscous diffusion into the inner core of the formed vortex

Assuming that the bulk velocity grows proportionally to t^{α} and that separation occurs from an edge of internal angle β, such that $\beta=0$ is a diaphragm and $\beta=90$ is a square corner. The flow velocity around the edge is given by $At^{\alpha}r^{\lambda-1}$ where r is the distance from the edge, A is a dimensional coefficient, and $\lambda=180/(360-\beta)$ (see Batchelor 1967, Sect. 6.5; Pullin 1978; Saffman 1992, Sect. 8.5). Then, from dimensional arguments, based on the fact that A and t are the only dimensional quantities available, the typical length size of the vortex increases in time as t^n, with $n=(1+\alpha)/(2-\lambda)$, and the vortex intensity (i.e. circulation), grows like t^{2n-1}.

These estimates allow evaluating the force acting on the surrounding walls due to the vortex formation, which turns out to be proportional to t^{2n-2} (see Sect. 2.6). Alternatively, they allow evaluating the flow corresponding to a given force or pressure difference. For example, in an orifice with zero internal angle, $\beta=0$ ($\lambda=1/2$), a flow time-profile t^{α} associates with a force going as $t^{(4\alpha-2)/3}$; this shows that a flow that increases faster than square root growth ($\alpha<1/2$) requires an unrealistic, theoretically infinite effort.

This typical roll-up phenomenon can be disturbed during its development when the shear layer is particularly thin. A curved shear layer is in fact subjected to an intrinsic instability that gives rise to the birth of wavy disturbance and the subsequent roll-up of multiple small double spirals along the shear layer itself (Pullin 1978; Luchini and Tognaccini 2002; Pedrizzetti 2010). This instability has the origin in the Kelvin-Helmholtz instability for an infinitely thin vortex sheet (Batchelor 1967, Sect. 7.1) and is more effective if the vortex layer is thinner. The thickness of

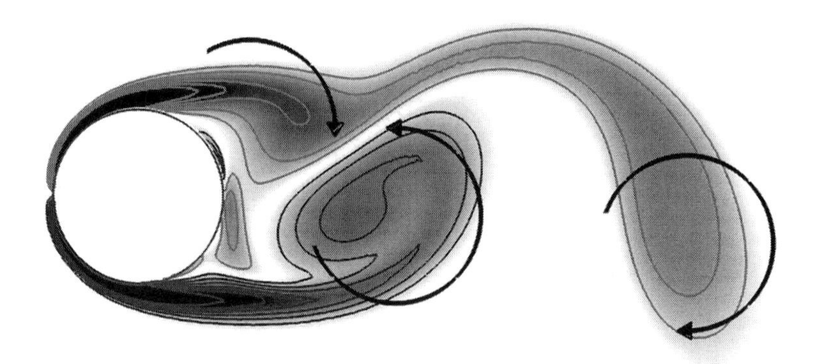

Fig. 2.5 Formation of vortices behind a circular cylinder. Oppositely rotating vortices separate from the two sides of the body in an alternating sequence. The previously separated clockwise vortex detached from the upper wall translated downstream, a counter-clockwise vortex has been formed from the lower wall, and a novel clockwise vortex is under formation from the wall above

the shear layer is related to the boundary layer thickness prior to separation, as given by Eqs. 1.16 or 1.17. In general, a separating shear layer is relatively thin in large vessels, especially when it detaches in a sharp enlargement after a short converging section, where the boundary layer is kept attached by the spatially-accelerating flow. This is the case of the trailing edge of heart valves.

The vortex formation from a smooth surface is similarly described with emphasis on a few additional elements. First, the actual position of separation depends on the local flow structure as it cannot be preliminarily identified and may even change in time (Pedrizzetti 1996). Furthermore, the separation from a smooth surface is inevitably accompanied by a more direct interaction between the forming vortex and the nearby wall when the viscous dissipation effects support the formation of smoother vortex structures.

One typical example of the external separation from the smooth surface of a bluff body is shown in Fig. 2.5. It features the formation of oppositely rotating vortices from the two sides of a circular cylinder. In this example, such vortices interact and influence the opposite separation process, and eventually produce a sequence of alternating vortices known as the *von Karman street* that is usually found behind bluff bodies (Panton 2005, Sect. 14.6). The development of alternating vortices is quite a common phenomenon when previously separated vortices influence vortex formation in nearby regions.

The internal separation, with subsequent formation of a vortex inside a vessel is generally a smoother phenomenon. This is due to the presence of confining walls that does not allow vortices to grow into large structures, keeps vortices more constrained within smaller scales, and is more influenced by viscous diffusion. Nevertheless, the presence of a vortex inside a vessel may disturb the entire flow. It has a blocking effect that locally deviates the streamlines, modifies the wall shear

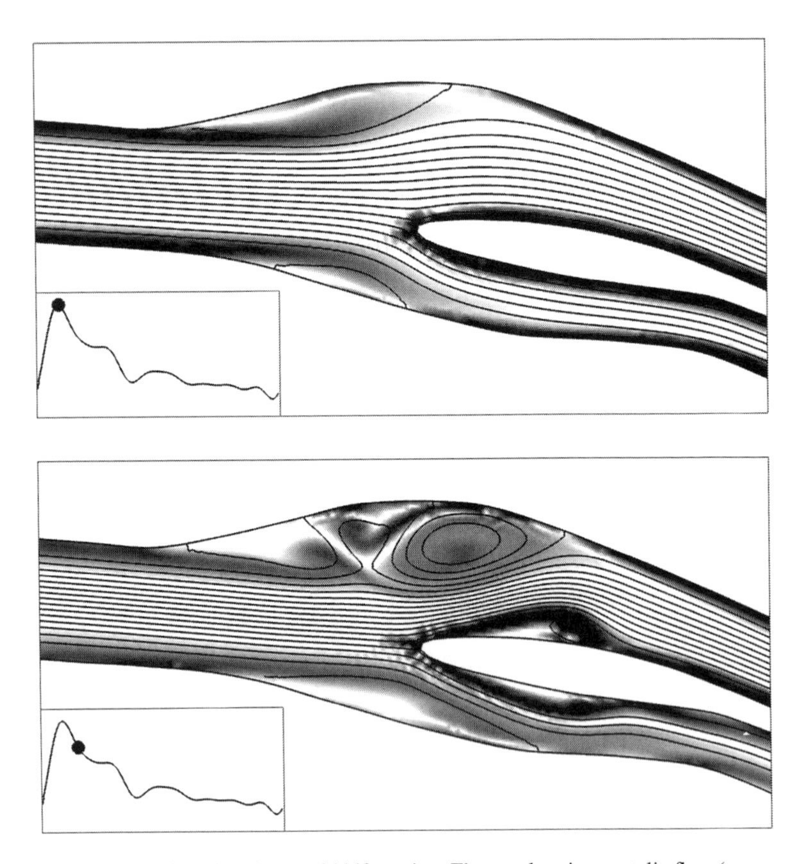

Fig. 2.6 Formation of vortices in carotid bifurcation. The accelerating systolic flow (*upper panel*, at peak systole) leads to a smooth boundary layer separation at the carotid bulb. After the peak (*lower panel*) the vortex just formed at the bulb interacts with the bulb boundary layer creating multiple small vortices, and gives rise to a vortex-induced secondary separation in the oppositely facing wall of the internal carotid artery. The same phenomena in a much weaker version are noticeable on the opposite side at the entrance in the external carotid artery

stress distribution, and possibly producing further separations. It changes the unsteady pressure drop and, in a branching duct, the relative flow-division in the daughter vessels. An example is given in Fig. 2.6 that depicts the vortex formation in the bulb of a carotid bifurcation. During the systolic acceleration, the boundary layer separates tangentially from the common carotid artery and develops a smooth roll-up within the bulb close to the nearby wall. During deceleration, the formed vortex locally affects the wall shear stress inside the bulb inducing opposite-sign wall vorticity. It has a blocking effect that deviates the streamlines at the entrance of the internal carotid artery into a faster jet. This produces secondary vortex-induced separation inside the internal carotid that eventually results in a secondary vortex formation and further small separations. A general analysis of the vortex formation process can be outlined when the flow enters from a small vessel into a large chamber as a jet with a as vortex. After the very initial roll-up phase, a measure of the length of such a jet is given by the product of velocity-time (Vt), where V is the velocity at the opening and t is the time. In this case, a dimensionless parameter, *vortex formation time* (VFT), was introduced by Gharib et al. in 1998, as the ratio of the jet length to the diameter of the opening nozzle, D:

$$VFT = \frac{V \times t}{D} \tag{2.7}$$

The vortex formation time represents a dimensionless number that characterizes the progression of vortex formation. It allows effective description of the vortex formation processes as occur under different conditions. The VFT also corresponds to the dimensionless measure of the vortex strength, the circulation Γ, normalized with VD. The definition (2.7) can be extended to the case when either V or D is time-varying, by integration of the ratio V/D during the period of vortex formation. The generality of the formation time concept allows uncovering general properties of the vortex formation. These characteristics will be discussed subsequently.

2.5 Three-Dimensional Vortex Formation

The vortex formation process described in the previous chapter is given in terms of two-dimensional images. They allow an immediate and intuitive understanding of the fundamental phenomena. The two-dimensional description implicitly treats the vorticity as a scalar quantity with a single component that is perpendicular to the two-dimensional plane of motion. To understand the three-dimensional nature of the vortex formation, it should be noted that vorticity is a three-dimensional vector. Vorticity is a solenoidal vector field, with zero divergence as dictated by the condition (2.4). It means that vorticity behaves like an incompressible flow. *Vortex lines*, lines everywhere tangent to the vorticity vector, must be continuous and cannot originate or terminate in the flow.

A consequence of the continuity of vortex lines is the concept of *vortex tube*, sometime referred to as the *vortex filament* when the tube is thin enough. A vortex tube is a thick collection of vortex lines; a tube whose lateral surface is made of vortex lines. The solenoidal nature of vorticity imposes that a vortex tube maintains its individuality during flow evolution. It is transported by the local flow, deformed and stretched by the velocity gradients, and enlarges because of vorticity diffusion while it maintains its individuality. When a tube approaches another tube, the nearby vortex lines belonging to different tubes begin to wrap one around the other, with a strong local stretching that is ultimately smoothed out by viscosity. In such "close encounters" a tube may fuse with other tubes, filaments, or vortex lines.

Another significant concept related to the three-dimensionality of vortices is the *self-induced velocity*. A two-dimensional vortex is actually a rectilinear vortex tube that does not vary along the third dimension where the corresponding velocity field is a rotation on and around the tube. Generally, a three-dimensional vortex tube is not rectilinear and presents a curvature that may change along its length. However, because of the relationship between velocity and vorticity, a curved vortex tube corresponds to a velocity field due to rotation around the tube plus a translation, or self-induced velocity of the tube itself.

Assuming that the rotational velocity around the curved vortex tube is such that the velocity is externally upward and internally downward; this means that every small portion in the curved tube also pushes downward the nearby elements with an overall downward translation. The self-induced velocity of a curved vortex filament is (Saffman 1992, Sect. 11.1) proportional to the vortex circulation Γ, its curvature $1/R$, where R is the radius of curvature, and also weakly influenced by the ratio R/c where c is transversal size of tube. The velocity here has an intensity of $\Gamma/4\pi R \log(R/c)$ and is directed perpendicular to the plane that locally contains the filament. Therefore, the tighter is the curve, the higher is the self-induced velocity.

The simplest type of three-dimensional vortex forms from a circular orifice in which, the forming vortex tube has the shape of a ring. Vortex rings are well known fluid structures (Shariff and Leonard 1992) that are typically generated using a piston-cylinder apparatus. A vortex ring is a stable vortex structure with axial symmetry. Because of their stability, vortex rings are frequently found in nature.

Figure 2.7 shows an instant corresponding to the formation process of a vortex ring through a circular orifice. The vorticity distribution on a transversal section (left panel) shows the shear layer separating from the orifice that eventually rolls-up into the jet head. The vortex core corresponding to the vortex ring is shown (right panel) to emphasize the ring shape of the three-dimensional vortex. In general, effective delineation of a vortex boundary in 3D is not trivial. This is usually easier in two-dimensional systems when the entire vorticity field can be shown in color scale on the picture plane, and the different elements of the vortex structure are immediately recognized (see the left panel of Fig. 2.7).

The definition of a vortex structure is a critical issue in three-dimensional flows. A generally accepted definition of a vortex is still lacking. The level of vorticity usually is not sufficient because the highest magnitudes are typically found in the separating vortex layer, or in the boundary layer. According to one definition, a vortex is a region where vorticity levels are higher than in the surrounding, where the fluid is subjected to a centripetal acceleration. The most accepted technique to identify a vortex structure is called λ_2 method introduced in by Jeong and Hussain (1995). They defined the vortex boundary by constant level of a scalar quantity, λ_2, obtained from the properties of velocity gradient.[1] In brief, this quantity identifies the regions where fluid pressure is minimum considering only the pressure gradients imputable to vorticity, and leaving aside the influence of the irrotational part of

[1] The scalar λ_2 is the intermediate eigenvector of the tensor build by the sum $D^2+\Omega^2$, where D and Ω are the rate of deformation and rotation tensors, respectively.

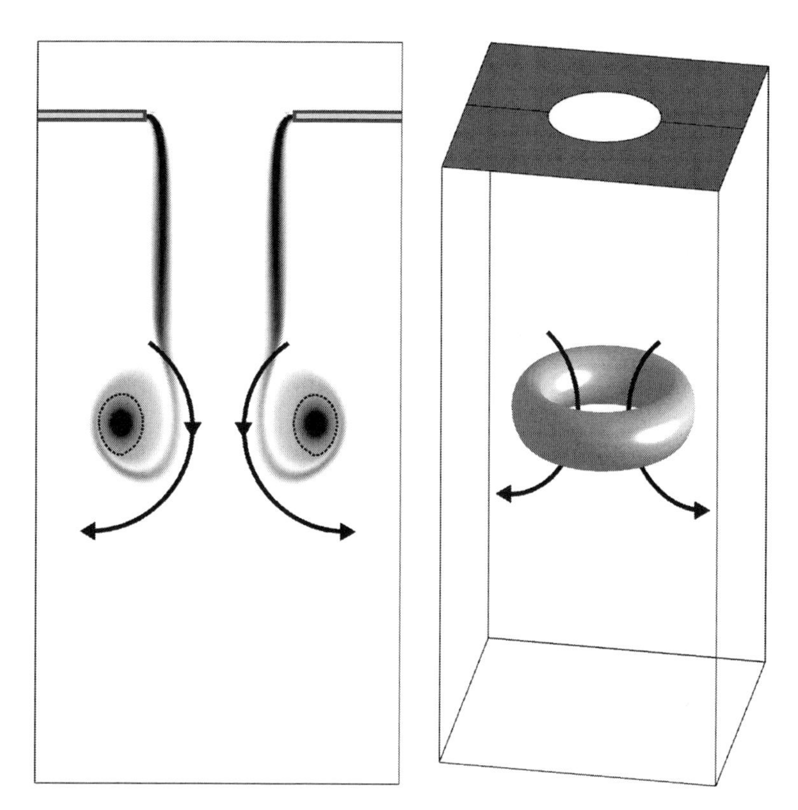

Fig. 2.7 Initial stages of vortex ring formation from a circular shape orifice. *Left panel*: distribution of vorticity on a transversal cross-cut; the vortex core is indicated with a *dashed line*. *Right panel*: three-dimensional view of the vortex ring corresponding to the core of the forming vortex visualized by the λ_2 method (see text). The shear layer separating from the edge rolls-up into a vortex ring that corresponds to the jet head. The vortex has a curvature, and the self-induced velocity is directed downstream and adds on top of the background velocity

the velocity. Therefore, the scalar λ_2 takes minimum values at the center of a vortex where pressure is low because centrifugal forces push the flow away. This technique has been successful in several applications, ranging from simple vortex flows to turbulence, which allows extracting and visualizing the coherent vortex structures.

A vortex ring presents a self-induced velocity proportional to circulation and curvature. Such a self-induced velocity gives rise to a peculiar limiting process of three-dimensional vortex formation that was first reported by Gharib et al. in 1998. During its formation, the vortex ring continuously is being fed by the rolling-up shear layer separating from the orifice edge. Therefore, its circulation grows and the self-induced vortex translation velocity increases. The self-induced translation velocity of the vortex ring rises until it exceeds the velocity of the separating shear layer. At this point,

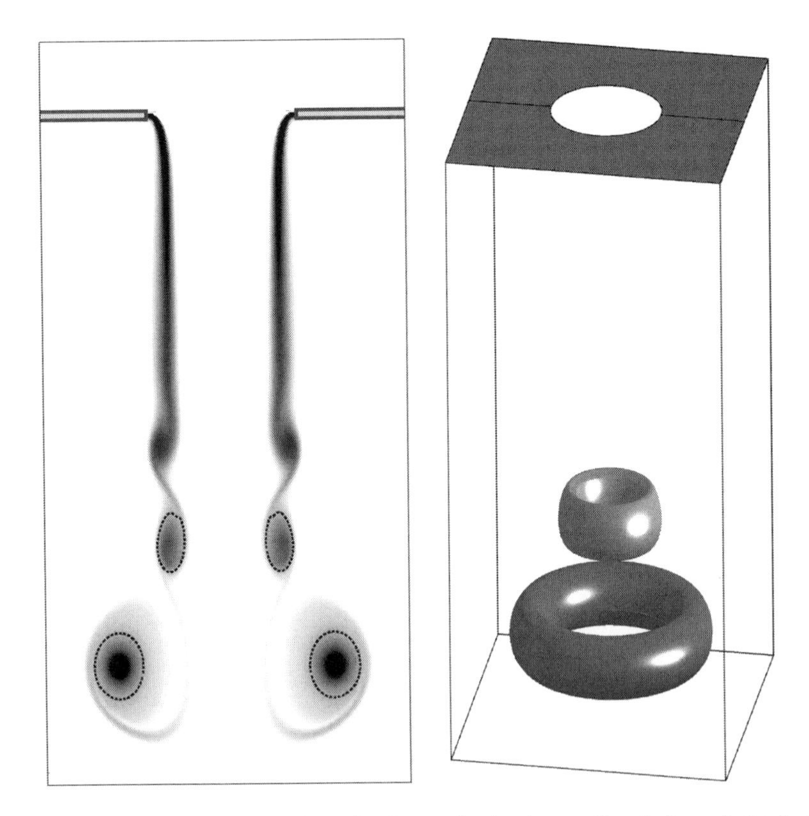

Fig. 2.8 Later stages of vortex ring formation from a circular shape orifice. *Left panel*: distribution of vorticity on a transversal cross-cut; vortex cores are indicated with dashed lines. *Right panel*: three-dimensional view of the vortex rings corresponding to the core of the forming vortices. The primary vortex has grown until its self-induced velocity has become larger than that of the shear layer behind, afterward the principal vortex escaped downstream and the shear layer produces smaller vortices in its wake

the primary vortex detaches from the layer behind with a phenomenon known as *pinch-off*. At that time, the newly separated vorticity cannot reach the escaped vortex and eventually rolls-up in its wake. In one sentence, *vortex ring pinch-off occurs when the velocity of the trailing jet falls below the celerity of the leading vortex ring* (Dabiri 2009).

The timing of this limiting process has been found to be well described in terms of the vortex formation time parameter defined in Eq. 2.7. Once it reaches to a critical value around 4, the vortex ring would be pinched-off. The VFT at which the vortex is pinched-off is called *vortex formation number*. Above this limit, the vortex ring cannot grow as a unique structure, and multiple trailing vortices develop in its wake. Indeed, the self-induced velocity of a curved vortex is proportional to Γ/D and the vortex formation number can be interpreted as the ratio between the velocity of the vortex ring and that of the shear layer. One example of the vortex ring formation process for a VFT larger than the critical value is shown in Fig. 2.8.

The case of vortex ring formation through a circular nozzle represents a preliminary conceptual basis for the interpretation of the more complex phenomena involved in the three-dimensional vortex formation from arbitrary-shaped nozzles. In an orifice with slender shape, the opening has a variable curvature and the separating vortex filament initially presents a variable curvature along its axis. These differences result in at least two potential effects in sequence. First, the self-induced velocity which is proportional to the curvature will be different along the vortex filament and will progressively deform it. Second, the vortex formation time depends on the local curvature and the vortex will reach the limiting critical value at different times along the different portions of the filament. These phenomena result in the deformation, and eventually, three-dimensional metamorphoses of the vortex structure separating from a non-circular orifice. Furthermore, once the vortex loses its stable ring-shape, the three-dimensional interactions lead to progressive destruction of the vortex into smaller elements, which in turn deform into even smaller ones, until they are dissipated due to the viscous effects. One exemplary case of the three-dimensional vortex formation from a slender orifice is shown in Fig. 2.9 (Domenichini 2011).

The three-dimensional vortex formation from smooth surfaces introduces additional elements of uncertainty, with variable separation lines, local roll-up, and formation of unstable three-dimensional vortex tubes, that does not allow drawing a simple unitary picture. In general, vortex formation from irregular-shape nozzles or uneven boundaries naturally leads to unstable vortex structures that are rapidly dissipated.

2.6 Energy Loss and Force of Vortex Formation

The previous sections in this chapter have evidenced how vortex formation process dramatically influences the flow motion. In addition to this, vortex formation also leads to generation of dynamical actions on the surrounding structures and energy loss in the flow. Consider an obstacle that partially obstructs the otherwise free-flowing of the fluid inside a vessel, as shown in Fig. 2.10. In the absence of any vortex formation, pressure would change as dictated by the Bernoulli balance. However, development of a vortex provokes an additional pressure drop (or energy loss) due to the transformation of energy into vortex inertia.

We evaluate the pressure drop in a simple configuration of a rectilinear duct with a diaphragm inside, as sketched in Fig. 2.11. This example allows defining the pressure drop due to vortex formation only, regardless of the transformation between kinetic energy and pressure caused by variation of the duct size. The evaluation of pressure loss in a partially obstructed straight vessel can be performed based on the equation of motion. Writing the equations along the streamlines and summing-up for all stream-tubes connecting the inlet to the outlet, the pressure drop between the inlet and outlet section of equal size can eventually be expressed as:

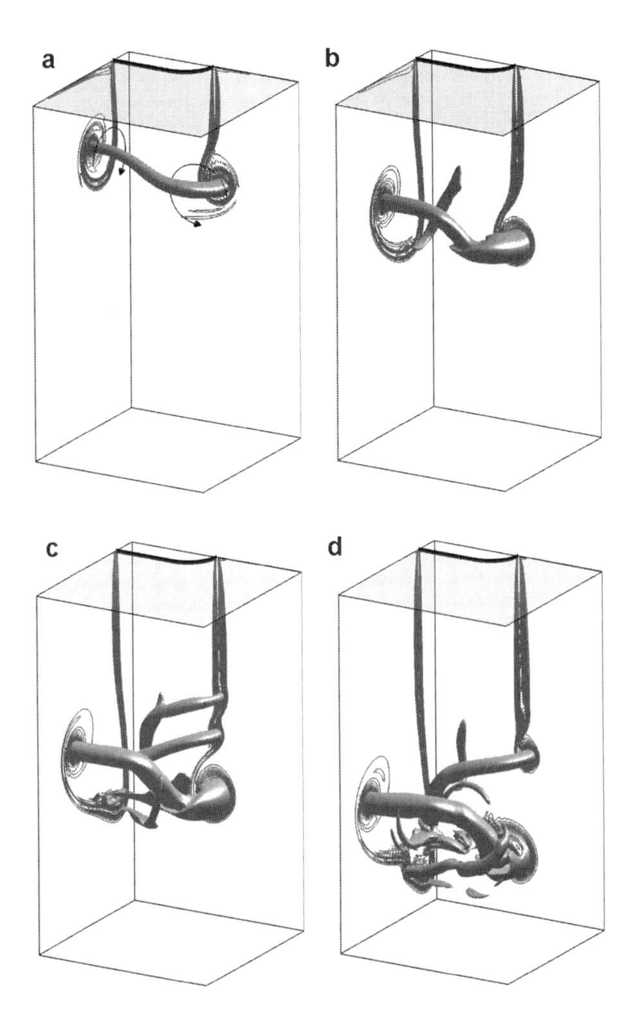

Fig. 2.9 Three-dimensional vortex formation from a slender orifice, made of two half circles connected by straight segments. The vortex formation time are (**a**) $VFT = 2.5$, (**b**) $VFT = 3.75$, (**c**) $VFT = 5$, (**d**) $VFT = 6.25$ from. One quarter of the entire space is shown for graphic clarity; the vorticity contours are reported on the side planes to help understanding the three-dimensional arrangement of the principal vortex filaments. In the initial phase, the formed vortex loop presents a variable curvature and deforms because of the higher self-induced translation speed in the more curved parts (**a**); such a deformation leads to further changes in the three-dimensional curvature and further deformations (**b**). Later on, the vortex reaches the limiting formation phase behind the circular part of the orifice only, where smaller vortices appear (**c**). Afterwards the vortex structure loses its individuality and becomes a set of entangled three-dimensional elements (**d**) that rapidly dissipate for viscous stresses

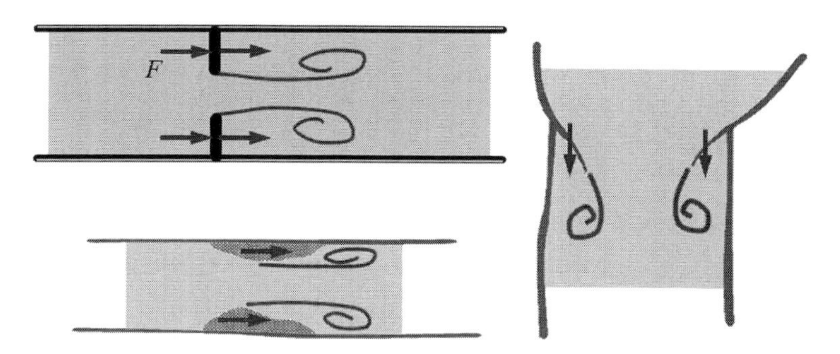

Fig. 2.10 A vessel that presents an obstacle partially obstructing the fluid flow, may give rise to vortex formation that generates a force on the obstacle. The picture depicts example of a duct with a diaphragm (*top-left*), a vessel with a stenosis (*bottom-left*), and a trans-valvular flow (*right*)

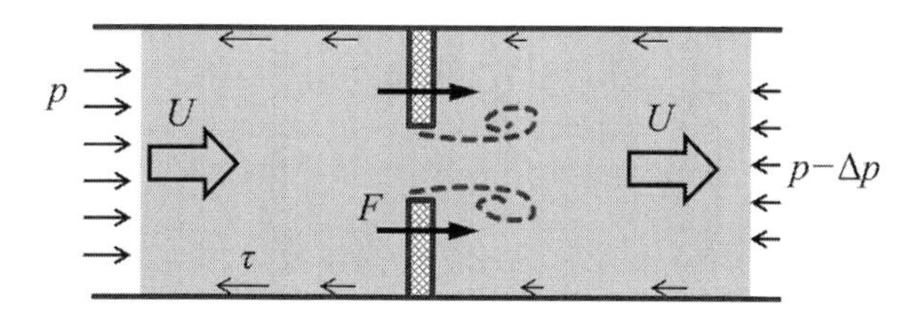

Fig. 2.11 A schematic portion of a straight duct with an obstacle. The process of vortex formation leads to an irreversible pressure drop Δp. This pressure drop can be estimated from global balances without the need to evaluate local details of the flow field

$$\Delta p = \rho L \frac{dU}{dt} + \frac{\rho}{A}\frac{dI}{dt} \tag{2.8}$$

where all the explicit viscous dissipative effects were neglected in this brief tract. Both terms on the right side were derived from the inertial part of the Bernoulli equation integrated over stream-tubes. The vortex impulse is defined as, I, a vector quantity (Saffman 1992, Sect. 3.2)

$$I = \frac{1}{2}\int \omega \times x dV \tag{2.9}$$

of which only the longitudinal component enters in (2.8). The apparently complicated definition of impulse (2.9) can be made more explicit when considering formation of an individual vortex that grows as $\Gamma(t)$ through the constriction area A_0.

The impulse is proportionate to the circulation multiplied by the area surrounded by the vortex $I(t) \sim A_0\Gamma(t)$; for example the impulse of a vortex ring of radius R is $I = \pi R^2\Gamma$, and the pressure drop simplifies as (see also Saffman 1992, Sect. 3.8)

$$\Delta p = \rho L \frac{dU}{dt} + \rho \frac{A_0}{A} \frac{d\Gamma}{dt} \tag{2.10}$$

The pressure loss caused by vortex formation, expressed by the second term of Eqs. 2.8 or 2.10, is an inertial effect. It is a consequence of the adherence of the fluid on the wall because of viscosity. In summary, vortex formation gives an irreversible transformation of energy into inertia, due to viscous adherence, but independent from the value of viscosity.

In addition to this, the presence of an obstruction in an otherwise rectilinear duct deviates the flow and provokes the development of a force on the obstacle whose strength increases during vortex formation process. This force results from the pressure difference between the front and back faces of the obstacle. The high pressure in the front face is due to the flow impacts. The dramatically low pressure on the back face is mostly a consequence of the vortex generation, and is due to the sharp pressure drop across the separating shear layer that connects the wall to the growing vortex. The total force exerted by the fluid on the obstacle can be evaluated by integrating the value of pressure and shear stresses all over its solid surface. This direct calculation is often not feasible because it requires a very accurate knowledge of the fluid properties adjacent to the interface with the solid boundary. An alternative approach is to use the integral balance of momentum (the product of mass and velocity integrated over the entire volume). This method states that the rate of change of momentum within any arbitrary region of fluid can only be imputable to the forces that act on the fluid inside that region (Panton 2005, Sect. 5.14). This is based on Newton's second law, expressed in integral form for an entire region:

$$\frac{dM}{dt} + M_{flux} = G + P \tag{2.11}$$

where the left side contains the rate of change of momentum dM/dt and flux of momentum M_{flux} across the open boundaries. On the right hand side, G indicates the volume forces, like gravity, and the symbol P represents the forces acting on all the surfaces confining the selected region. These terms consider the contribution of pressure difference at the open ends of the region and the total stresses acting over the intersection of the fluid and obstacle, which is equal and opposite to the unknown force made by the flow over the obstacle.

Considering the same simple duct with constant cross-section sketched in Fig. 2.11, we discuss on evaluating the longitudinal force, F, exerted over the solid obstacle due to the vortex formation process. In this case, the total flux of momentum is zero because the same amount that enters from the inlet exits from the outlet, and the volume force G due to gravity is the static weight of the fluid that can be ignored. The longitudinal balance (2.11) thus expresses a dynamic equilibrium between the change of momentum, dM/dt, and the surface forces, P. The product of the fluid density and the velocity gives the momentum M over the whole volume. In a duct, the whole volume can be spanned by a sequence of cross-sectional slices. The integrated velocity is the total fluid discharge, $Q = U \times A$, across that section. When the lateral walls do not move or their velocity is negligible, the discharge does not vary along the length of the duct, and the rate of change of momentum is proportional to the rate of change of the discharge. This can be mathematically expressed as $\rho \times L \times dQ/dt$, where ρ is the fluid density, and L is length of the considered portion of the duct. The surface force, P, presents two different contributions. First, the forces acting on the open ends: the forward pushing pressure at the inlet section and the backward pushing pressure at the outlet, which sum up to $\Delta p \times A$. Second, the total longitudinal force due to the entire solid boundaries, $-F$. This includes both actual force across the contour of the obstacle and the viscous shear stress on the lateral surfaces that are negligible along the short tracks. After summation, Eq. 2.11 becomes

$$\rho L \frac{dQ}{dt} = \Delta p A - F \tag{2.12}$$

Substitution of the expression for the pressure drop that was previously evaluated in Eq. 2.8 shows that the force generated by vortex formation over an obstacle can be primarily expressed as:

$$F = \rho \frac{dI}{dt} \tag{2.13}$$

neglecting the additional contributions that may come from viscous loss. In general, when the ducts or chambers possess a more complicated geometry with moving boundaries, the equation for the developed force involves other terms as well. However, these additional terms, do not relate directly to the vortex formation, and can be typically estimated based on the geometry and average flow properties.

In the single vortex with circulation $\Gamma(t)$ across the area A_0, when the pressure drop is expressed by Eq. 2.10, the force simplifies as:

$$F = \rho A_0 \frac{d\Gamma}{dt} \tag{2.14}$$

This expression allows calculation of the force associated with the simple vortex formation when the intensity growth of the vortex is somehow obtainable. For example, the initial strength of a starting vortex can be evaluated from general vortex formation concepts, like those introduced in Sect. 2.4. Therefore, when a flow accelerates across a sharp orifice, the resulting force during the starting phase turns out to be proportional to $\rho A_0 (U^2/t)^{2/3}$. When the leading vortex is far downstream, the separating shear layer with the intensity of U, translates straight downstream with a velocity proportional to U. In this case, the emitted circulation rate is $d\Gamma/dt$ proportional to U^2, and the force is about $\rho A_0 U^2$.

The vortex formation phenomenon generates an additional force over the obstacle that is proportional to the orifice area and the rate of growth of the vortex strength. This *force generation* may be beneficial or detrimental depending on the location; for example, force generation due to transmitral vortex is beneficial to the left ventricular function while vortex formation downstream a stenosis can be detrimental.

2.7 Vortex Interactions

When two or more vortices come nearby each other, they likely interact in an intense and irreversible manner. The interaction of vortices involves many different and complicated phenomena, which we have outlined the most important ones here.

A 2D vortex is associated with a rotating flow whose velocity is proportional to the vortex circulation, Γ, and inversely proportional to the distance, r, from the vortex center: $v = \Gamma/2\pi r$. Two vortices that come in close encounter reciprocally induce a rotational velocity to each other. When these vortices have the same sign, they rotate together around each other. The differential velocity within each individual vortex deforms it and makes them winding up over each other, and eventually merge into a single larger one equivalent to the sum of them. This process is associated with little energy dissipation. On the contrary, two vortices with opposite circulation, make a vortex pair, which translate together due to the self-induced velocity along a straight or curved path depending on the relative strengths. Again, the differential velocity inside each single vortex produces the winding up of one's vorticity strip on the other. However, such vorticity strips are of opposite sign and do not merge rather they annihilate each other and reduce the individual vortices' strength.

The close encounter of three-dimensional vortex loops begins with the local interaction between the closest tubular elements of the two vortices that is initially two-dimensional. A close encounter between tubular elements with the same sign is an extremely rare event because the overall self-induced velocity of the corresponding vortex loops tend to separate the vortices. Thus, three-dimensional interaction begins prevalently between two oppositely rotating portions of a vortex tube. One example of the interaction between two identical vortex rings is shown in Fig. 2.12. Initially, the local interaction is approximating the two-dimensional process described above; the nearby oppositely rotating tubular elements induce the velocity to each other and try to translate away. This produces a local stretching of the three-dimensional vortex tube, a stretching that accelerates while the tubes come closer to each other and, in a non-symmetric case, would locally wind up on each other. The interacting structures develop increasingly small-scales until viscous dissipation becomes a dominant effect, at this point the *reconnection of vortex lines* occurs, and adjacent opposite vorticity is annihilated by dissipation and the vortex tubes tend to fuse one onto the other (Kida and Takaoka 1994).

The interaction between two identical vortices, like the ones shown in Fig. 2.12, may result into a complete vortex reconnection and a relatively simple vortex tube.

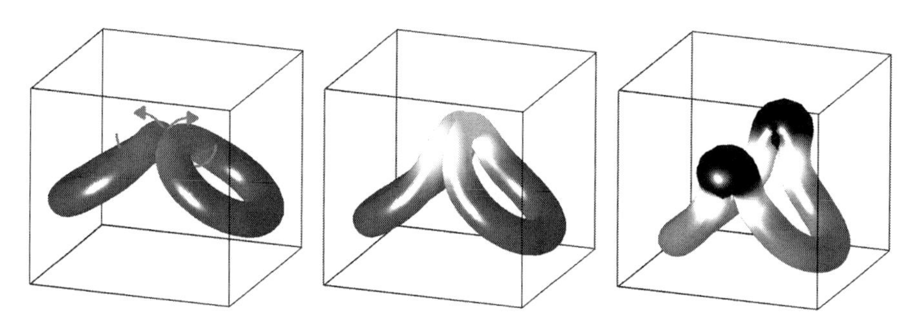

Fig. 2.12 Vortex reconnection and topological metamorphosis between two impacting vortex rings of equal circulation; the brightness of the filament indicates the strength of the corresponding vorticity. When oppositely rotating vortex tubes get close, they produce a local vortex stretching due to the self-induced velocity (from *left to central panels*). During stretching, the boundary between the vortices becomes locally sharper until the filaments fuse into each other due to viscous effect (from *central to right panels*). After vortex reconnection, a new structure is formed, typically with irregular geometry. This vortex is often unstable and short lived

More often, one vortex is stronger than the other one. In that case, only part of its tubular structure can reconnect with the other weaker vortex, and the incomplete reconnection gives rise to new vortices with a complex branched geometry. Typically, the vortex structure resulting from the fusion of previously interacting vortices, presents a very irregular geometry. An irregular three-dimensional vortex structure is overall unstable, tends to destroy itself, and is short lived. The more a vortex is regular, like a vortex ring, the more it remains coherent, and will last longer.

A special case of vortex interaction which particularly occurs in closed systems like cardiovascular vessels, is vortex *interaction with a nearby wall*. The vortex-wall interaction can be described as two different phenomena: the *irrotational interaction*, as a consequence of the wall impermeability; and the *viscous interaction* with the vorticity in the boundary layer. Here we discuss both cases:

First, an isolated vortex induces a rotary motion where streamlines are circular. Once such a vortex approaches an impermeable wall, the streamlines must deform to avoid crossing the boundary. With reference to Fig. 2.13 (left panel), the modification of the flow field that satisfies the impermeability condition can be immediately constructed based on symmetry considerations. It is the irrotational flow that would be induced by an *image vortex* with opposite circulation placed symmetrically below the wall. Such an image vortex results in a velocity perpendicular to the wall that is opposite to that of the real vortex, and thus ensures that the fluid does not penetrate into it. On the contrary, the tangential velocity has the same sign as the real vortex, and so, the velocity adjacent to the wall increases (*splash effect*). In addition, the image vortex induces a velocity to the real vortex that accelerates or decelerates (depending on the direction of the circulation) with respect to the background flow because of this *image effect*. For example, a clockwise vortex that just formed from a wall underneath is decelerated by the image below the same wall while accelerates when approaches a wall on the opposite side.

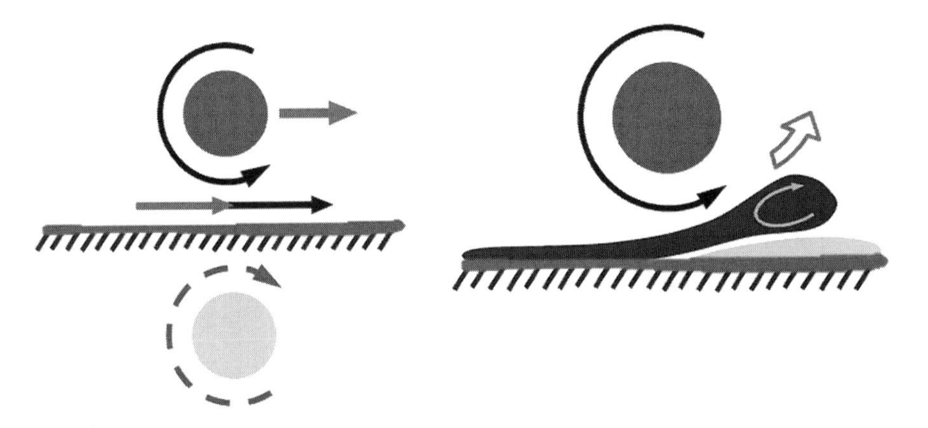

Fig. 2.13 The interaction of a vortex with the wall produces two separate effects. First (*left panel*), the condition of impermeability is satisfied by a distortion to the vortex-induced flow that is equivalent to having an opposite vortex placed symmetrically below the wall. The presence of such an "image" vortex increases the tangential velocity next to the wall, and induces a translation velocity to the otherwise still vortex. The second effect (*right panel*) is due to viscous adherence, where a boundary layer develops and eventually a vortex-induced separation occurs

Second, in addition to the image effect, a vortex near the wall also influences the development of the boundary layer because of the viscous adherence condition at such a wall. A vortex creates a local velocity gradient along the wall with acceleration followed by deceleration (or vice versa depending on the direction of rotation). This perturbation, as previously discussed in Sect. 2.3, may give rise to a vortex-induced separation of the boundary layer and to formation of secondary vortices as sketched in Fig. 2.13 (right panel).

When the vortex-boundary interaction applies to a tract of a three-dimensional vortex tube, it eventually affects the following three-dimensional dynamics: first, the image effect results in a local stretching and deformation of a vortex filament. Second, when the vortex gets closer, it eventually interacts directly with the vortex-induced vorticity distribution. This is an interaction between oppositely circulating vorticities that gives rise to the local wind-up of the wall vorticity around the approaching vortex, and to reconnection with its vortex lines. Eventually, the vortex crops by dissipation in the regions closer to the wall. This unbalances the three-dimensional vortex structures that tends to rapidly deform and develop small structures that are eventually dissipated.

2.8 A Mention to Turbulence

Let us briefly describe the realm of fully developed turbulence by further discussing the concept of interaction between three-dimensional vortices. The interaction between two vortices initially deforms the overall, *large scale* geometry of the vortex loops, and then, after sequences of reconnections, breaking of vortices and further deformations, it eventually transforms the original vorticity into several irregular small structures. Such *small scale* flow elements result in sharp velocity gradient, viscous friction, and rapid dissipation.

Now consider this vortex breakdown phenomenon in a condition where large vortices are continuously developed from the surrounding boundary. The resulting flow simultaneously contains these large structures along with the others at a variety of size scales. A measure of the complexity of such a flow can be obtained from the extent of such vortices, measured by the ratio of the largest scale, L, to the smallest one, indicated with η. When L is comparable to η, the flow is considered regular. For example, if the system continuously creates vortex loops of size L, the resulting flow is considered sequence of individual rings that decay as time proceeds since there are no vortices of a smaller size. On the opposite end, when L is much larger than η, the flow contains newly generated vortex loops of size L, previously generated vortices that broke down into smaller structures and a large number of interacting small vortices. The order of magnitude of this complexity can be estimated from the phenomenological theory of turbulence (based on Kolmogorov in 1941, reported, in Frisch 1995, Sect. 7.4)

$$\frac{L}{\eta} \approx Re^{3/4} \tag{2.15}$$

where Re is the *Reynolds number*

$$Re = \frac{UL}{\nu} \tag{2.16}$$

as previously introduced in (1.21). When Reynolds number is large enough, the flow shows fluctuations on velocity and vorticity over a wide range of scales and is classified as *turbulent flow*.

An increased friction between fluid elements and enhanced energy dissipation with respect to regular fluid motion characterize turbulence. In fact, the development of turbulence is the strategy used by the fluids to dissipate the excess of energy. When the fluid motion presents with a large density of energy (e.g. high velocity), the fluid may be unable to maintain the equilibrium between viscous dissipation and the external energy sources. In that case, it increases the particle paths by developing swirling motions with higher shear rate to increases viscous dissipation up to equilibrium. In fact, Reynolds number represents the ratio of the kinetic energy introduced in the large scales (proportional to ρU^2) to their ability to dissipate with shear stress, i.e. $\rho \nu U/L$. When Reynolds number increase above a certain threshold, the smaller scales develop to enhance dissipation. In other words, regular flow becomes unstable, and turbulence appears. Every perception of the flow presents with a *critical value of the Reynolds number* for development of turbulence. The value of the critical Reynolds number was mentioned in Sect. 1.6 to be about 2,300 in the case of steady flow in a circular vessel.

Turbulence enhances energy dissipation, which results in excessive energy consumption in the vasculature. The unpredictability of its chaotic fluctuations makes turbulent flows difficult to control, model, and predict. On the other hand, turbulence has several positive implications; first of all, it makes life possible by enhancing mixing and diffusion. While viscous diffusion is an extremely efficient mechanism to distribute substances at very small scales, turbulent dispersion dominates mixing at larger scales. For example, viscous diffusion length which grows proportionally to $\sqrt{t}$ (see Eq. 1.16), in water takes a few hundredth of a second to reach 1 mm, a few second for 1 cm, and over 11 hours for 1 m. On the contrary, the *accelerated turbulent dispersion* dominates the mixing and heat propagation at sufficiently larger scale, typically above a few millimeters. It is evident how turbulence is ubiquitous in nature, and how it ensures the mixing that is experienced in everyday life.

In general, the turbulence can be considered as a system of entangles and interacting vortex elements with disparate sizes, ranging from the large size generated by the boundaries, to the smaller size where the flow is smoothed out by viscous effects. These vortices are not clear individual structures like those discussed in previous sections. They are *turbulent eddies*, loosely defined as blob of vorticity of arbitrary shape coming from the breakdown of the large unstable vortices. Turbulence is thus a crowd of eddies that is stretched and twisted by the velocity field induced by vorticity field itself. Turbulence is a spatially complex distribution of vorticity that exhibits a wide and continuous distribution of scales, which advects itself in a chaotic manner.

The overall dynamics of turbulence is usually described in terms of *energy cascade* (Davidson 2004, Sects. 1.6 and 3.2). An external energy input (slope of a channel, a pumping pressure) pushes a fluid within its boundaries across an orifice, around an obstacle, or along an irregular vessel bend. The flow thus generates vortices whose sizes are comparable with that of the container. These vortices interact and produce smaller eddies that further interact and produce turbulent eddies with progressively smaller size. While eddies become smaller, velocity gradients are larger, and viscous shear stresses increasingly dissipate kinetic energy into heat. At the lower end of this energy cascade, very small eddies are entirely dissipated, and do not generate anything smaller. Thus, energy is injected into the turbulent flow at large scales, and is dissipated in the smallest scales of the flow due to viscous friction.

The most common strategy to tackle the problem of turbulence relies on statistical methods, defining a description of the average motion responsible for transport, and its fluctuations responsible for dispersion. However, turbulence is not a random process. It is actually a deterministic phenomenon; a turbulent flow is a solution of the mathematical equations governing fluid motion (the Navier-Stokes Eq. 1.14). The underlying deterministic nature of turbulence often emerges in the form of *coherent structures*, developing within an incoherent chaotic background. These coherent structures are principally the large scale vortices generated during vortex formation processes, possibly modified due to the presence of turbulence. Sometimes, they are vortices that develop due to the instability of a parallel flow like in boundary layer flows (Davidson 2004, Sect. 4.2.6). In some other cases, they emerge from coalescence of incoherent background vorticity or three-dimensional filaments. Coherent vortex structures typically hold most of the energy, and so, they represent the fundamental objects in the analysis of each specific turbulent flow.

In the cardiovascular system, turbulent flows are rarely encountered. The largest scales of motion achievable in the arterial network cannot exceed the vessel size, which is a few centimeters at most. The Reynolds number is normally well below 1,000, with the exception of the largest vessels. The flow in the ascending arota and, sometime, in the left ventricular cavity can reach values of the Reynolds number up to few thousands, just above the critical threshold in a short interval. The turbulence developed is *weak turbulence* with an energetic level that does not significantly influence the main dynamics and vortex formation processes. Such a weak turbulence simply increases the dissipation level through intense interactions between larger vortices. It should be remarked that the highest levels of turbulence, if any, in an unsteady pulsatile flow are recorded during the deceleration after the peak flow. In fact, although the instantaneous Reynolds number has decreased, the flow has been filled with the energy during the maximal velocity and has to dissipate such energy during deceleration. Deceleration enhances shear layer instability phenomena, and boundary layer separation, which both support turbulence.

Weak turbulence may develop during diastolic filling of the left ventricle when the trans mitral jet fills the heart. Additionally, the tri-leaflet geometry of the aortic valve provokes a rather complex three-dimensional vortex formation associated with the large Reynolds number (roughly from 3,000–8,000), which gives rise to interactions that produce small scales vorticity and weak turbulence. This phenomenon is even more enhanced in presence of mechanical valves.

References

Batchelor GK. An introduction to fluid dynamics. Cambridge: Cambridge University Press; 1967.

Dabiri JO. Optimal vortex formation as a unifying principle in biological propulsion. Annu Rev Fluid Mech. 2009;41:17–33.

Davidson PA. Turbulence. An introduction for scientists and engineers. Oxford: Oxford University Press; 2004.

Domenichini F. Impulsive vortex formation from slender orifices. J Fluid Mech. 2011;666: 506–20.

Frisch U. Turbulence. Cambridge: Cambridge University Press; 1995.

Gharib M, Rambod E, Shariff K. A universal time scale for vortex ring formation. J Fluid Mech. 1998;360:121–40.

Jeong J, Hussain F. On the identification of a vortex. J Fluid Mech. 1995;285:69–94.

Kida S, Takaoka M. Vortex reconnection. Annu Rev Fluid Mech. 1994;26:169–77.

Luchini P, Tognaccini R. The start-up vortex issuing from a semi-infinite flat plate. J Fluid Mech. 2002;455:175–93.

Lugt HJ. Vortex flow in nature and technology. New York: Wiley; 1983.

Panton RL. Incompressible flow. 3rd ed. Hoboken: Wiley; 2005.

Pedrizzetti G. Unsteady tube flow over an expansion. J Fluid Mech. 1996;310:89–111.

Pedrizzetti G. Vortex formation out of two-dimensional orifices. J Fluid Mech. 2010;655: 198–216.
Pullin DI. The large-scale structure of unsteady self-similar rolled-up vortex sheets. J Fluid Mech. 1978;88:401–30.
Saffman PG. Vortex dynamics. Cambridge: Cambridge University Press; 1992.
Shariff K, Leonard A. Vortex rings. Annu Rev Fluid Mech. 1992;24:235–79.
Wu JZ, Ma HY, Zhou MD. Vorticity and vortex dynamics. Berlin/Heidelberg: Springer; 2006.

Chapter 3
Vortex Formation in the Heart

Abstract The presence of vortical flow structures that develop inside different cardiac chambers is shown to correlate with functional status of heart, and significantly affects the cardiac pumping efficiency. In this chapter, formation of vortices at different locations inside the heart, and their physiological and clinical significance is discussed.

3.1 Mitral Valve and Transmitral Flow

3.1.1 Mitral Valve Functional Anatomy

Mitral valve (*valvula bicuspidalis*) is the left atrioventricular (AV) valve that relates left atrial chamber to the left ventricle. The valve consists of two triangular cusps primarily composed of elastin, collagen and glycosaminoglycans (GAGs). The forming components are organized in layers with a collagenous core (fibrosa) packed in between two elastin/GAG layers. These layers act as a continuous laminar structure, which are thought to exhibit transversely isotropic material properties (Chen et al. 2004). The valve's cusps are of unequal size, and are larger, thicker, and stronger than those of the tricuspid valve on the right side of the heart. The larger cusp, known as the **anterior cusp**, is positioned in front and to the right between the mitral orifice and the left ventricular outflow tract, while the smaller or **posterior cusp** is placed behind and to the left of the mitral orifice (Fig. 3.1). The cusps of the bicuspid valve are connected to the papillary muscles via the cord-like tendons that are called *chordae tendineae*, or heart strings. Chordae tendineae are roughly composed of 80% collagen, 20% elastin covered by endothelial cells.

The mitral annulus (*annulus fibrosus*) is described as a three-dimensional D-shaped structure that approximates a hyperbolic paraboloid, similar to a riding saddle where its peaks furthest from the LV apex (Fig. 3.1), located anteriorly and posteriorly, and its valleys located medially and laterally at the commissures (Levine et al. 1989; Bonser et al. 2011). This nonplanar, D-shaped cord-like, fibrous structure provides

A. Kheradvar, G. Pedrizzetti, *Vortex Formation in the Cardiovascular System*,

DOI 10.1007/978-1-4471-2288-3_3,

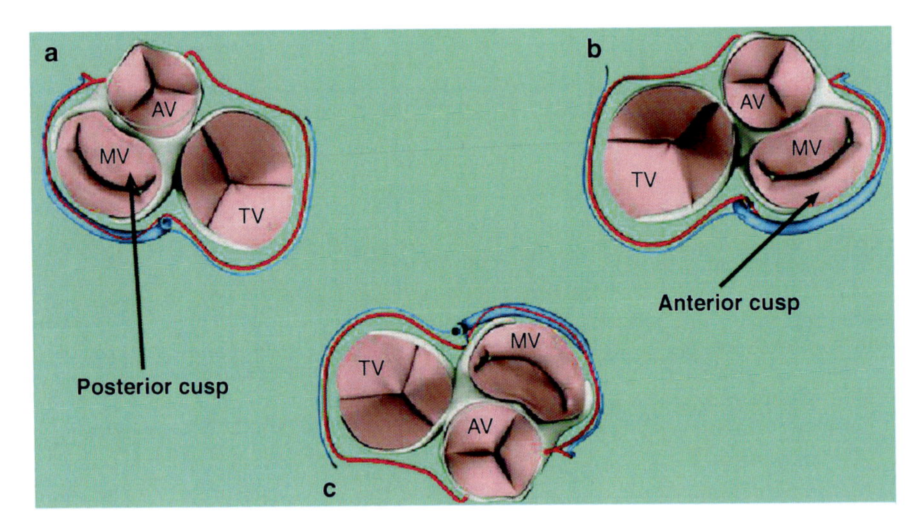

Fig. 3.1 Schematics of the mitral valve from (**a**) Surgeon's view; (**b**) transesophageal echocardiographic view; and (**c**) transthoracic parasternal view. The anterior and posterior cusps of the mitral valve are shown. *AV* aortic valve, *MV* mitral valve, *TV* tricuspid valve (Modified from Bonser et al. (2011))

the attachment of the mitral leaflets. The fibrous nature of the annulus provides a firm origin for the leaflets and ensures the leaflets' competency during the cardiac cycle. The mitral annulus is merged into anterior and posterior leaflets. The anterior side of the annulus is anatomically coupled to the aortic annulus. Surgeons commonly refer to this conjunction as the "*aortomitral curtain*", and cardiologists identify it as "*intervalvular fibrosa*". The posterior annulus is externally related to the myocardium at the left ventricular inflow region and internally to the left atrium where it merges with the base of the posterior mitral leaflet (McAlpine 1975; Berdajs et al. 2007). It is hypothesized that the dynamic, sphincter-like function of the annulus follows the active atrial contraction during systole (Berdajs et al. 2007).

3.1.2 Transmitral Flow

Diastole begins with isovolumic relaxation phase, a complex energy releasing period during which the myocardial contractile elements are deactivated and return to their original length. This geometric change in contractile elements results in sudden pressure drop inside the left ventricle (LV). The immediate response of the LV to this sudden pressure drop is the opening of mitral valve. Accordingly, both diastolic suction and initial ventricular pressure drop actively contribute to early ventricular filling. The rapid pressure drop due to isovolumic relaxation is a phenomenon that guarantees adequate ventricular filling volume at low pressure. As the pressure continues to decline in the LV, the transmitral flow follows from left atrium (LA) to LV. The mitral valve tends to close when the transmitral pressure gradient reverses and the atrio-ventricular flow decelerates. However, prior to complete closure of the valve, atrial contraction phase begins, and the left atrial pressure increases to maintain the pressure gradient. At this moment, blood is ejected from LA to the LV due to the higher atrial pressure.

Normal transmitral flow is usually laminar and relatively low in velocity (usually less than 100 cm/s). Pulsed-wave (PW) Doppler is currently the preferred imaging modality to quantify localized velocity of transmitral flow as it measures the blood flow velocity within a small area at a specified tissue depth. When flow across the mitral valve is assessed with PW Doppler, two filling waves are characteristically observed; the first wave (E-wave) represents early diastolic flow velocity, which is caused by the sustained myocardial relaxation, and the second wave (A-wave) characterizes the active filling due atrial contraction (atrial systole). Typically, E-wave velocity is slightly greater than the A-wave's.

3.1.3 Transmitral Vortex Formation

The presence of vortical flow structures that develop along with a strong propulsive transmitral jet was initially contemplated through *in vitro* studies of the ventricular flow (Bellhouse 1972; Reul et al. 1981) and subsequently confirmed by analyses based on color Doppler mapping (Kim et al. 1994) and Magnetic Resonance Imaging (MRI) (Kim et al. 1995; Kilner et al. 2000). The early transmitral flow is considered a rapidly starting jet that forms a vortex ring from the boundary layers at the distal tip of the mitral valve leaflets. The shear layer that emerges from the mitral valve leaflets rolls up into an asymmetric vortex ring (Kheradvar et al. 2011). This asymmetric flow structure (Fig. 3.2) is mainly due to the unbalanced shape of the mitral valve whose anterior leaflet is larger than the posterior leaflet, and the uneven interaction of the forming vortex with ventricular walls. During the formation process, the vortex propagates away from the mitral valve and entrains the ambient fluid inside the LV until it is pinched-off from the transmitral jet (Fig. 3.2). The vortex ring also deforms when it is being detached from the mitral leaflets due to the non-uniformity of the ring thickness as it propagates toward the apex. This deformation is not only related to disparity of the leaflet lengths but also may be attributed to the inhomogeneous pressure gradient inside the LV and the interaction with the LV wall. The details of the vortex ring formation process, up to its deformation and breakdown during diastole has been described by numerical simulation in an idealized ventricular shape in (Domenichini et al. 2005).

It is shown that the physical characteristics of these vortices provide additional information with respect to physiological events related to diastolic functionality (Kheradvar et al. 2007, 2008; Kheradvar and Gharib 2007, 2009). The process of vortex formation is further described by the vortex formation time index, a non-dimensional measure of early filling period (Kheradvar and Gharib 2009) that is equivalent to the stroke ratio of the transmitral jet:

$$VFT = \frac{\bar{U}}{\bar{D}}T \tag{3.1}$$

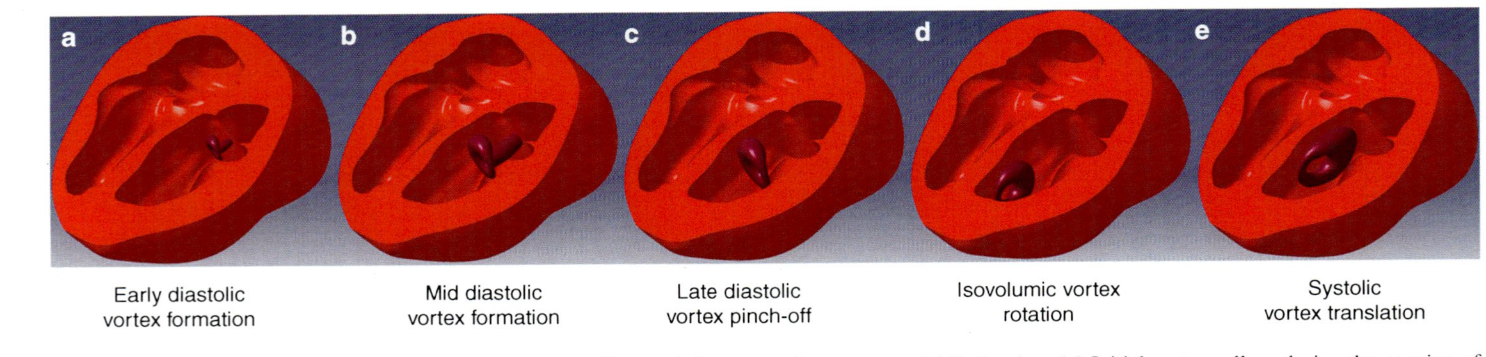

Fig. 3.2 Schematic of the vortex formation process in a cardiac cycle in a normal reconstructed LV chamber. (**a**) Initial vortex roll-up during the opening of the mitral valve; (**b**) mid diastolic vortex formation; the vortex is still growing while attached to the transmitral jet; (**c**) vortex pinches-off from transmitral jet during late diastole in a normal heart; (**d**) the pinched-off vortex keeps rotating during isovolumic contraction phase; (**e**) the vortex that formed and pinched-off during diastole translates toward aortic valve during systole (Modified from Kheradvar et al. (2011))

where "$\bar{U}$" is the mean velocity of the starting jet, T is the duration of rapid filling phase (E-wave) and $\bar{D}$ is the average open diameter of the mitral orifice. Following the discussion previously given in Sect. 2.5, it has been shown that by increasing the starting jet's stroke ratio or vortex formation time greater than the range of 3.5–5.5, no additional energy or circulation enters the leading vortex and the remaining fluid in the pulse ejects as the trailing jet (Gharib et al. 1998). After this stage, the vortex is said to have *pinched off* from its generating jet, and the size of the leading vortex does no longer increase. The vortex formation time at which the vortex is pinched off is called *vortex formation number* that is a measure of optimal ejection (Gharib et al. 1998). The importance of this parameter lies in the fact that in a jet with VFT exceeding the vortex formation number range (~3.5–5.5), the remaining fluid in the pulse ejects as the trailing jet. Thus, VFT index defines the fate of the vortex between two possible scenarios of formation: either a single vortex ring, or a leading vortex ring with trailing jets. A single vortex ring is produced when VFT is less than or within the vortex formation number range, while a vortex ring with trailing jets is produced in case of larger VFT such as the case of mitral stenosis (Kheradvar et al 2011). The critical role of vortex formation rests on the relative contribution of the leading vortex ring and the trailing jet to the thrust supplied to the flow (Krueger and Gharib 2003).

The leading vortex transfers extra momentum from LA to LV contributing to efficient blood transport toward the aorta. The additional sources of momentum-transfer either come from the added mass effect (Krueger and Gharib 2003) in which the streamlines would act as a boundary that drives the ambient fluid into motion when the vortex is being formed, or result from fluid entrainment inside the isolated transmitral vortex bubble (Dabiri and Gharib 2004). Accordingly, some of the residual blood inside the ventricle (ahead of the transmitral jet) is accelerated ahead while the transmitral jet is being initiated. At the same time, some ambient fluid must be brought in behind the vortex ring to preserve the continuity of the flow (Kheradvar et al. 2011).

Although there is an overall agreement on existence of transmitral vortex formation, no firm agreement exists on shape and rotational direction of vortex structures during normal transmitral flow. Recent works report that in a normal heart, due to asymmetry of the leaflets, the transmitral vortex tends to be more asymmetric than symmetric (Kheradvar et al. 2010; Pedrizzetti et al. 2010). Based on the current *in-vitro* studies, changes in mitral leaflet's length could vastly affect the transmitral vortex formation possibly due to altered flow-wall interaction. Similar results have been reported in different conditions involving flow-wall interactions (Kheradvar et al. 2007; Shariff and Leonard 1992). Kheradvar and Falahatpisheh (2012) showed that the transmitral vortex formation can be influenced by the saddle annulus dynamics, and the angle of valve opening. They showed that a symmetric, donut-shape vortex ring is formed along with the transmitral jet downstream a bileaflet mitral prosthesis with leaflet height of 11 mm. However, increasing the leaflet's length to 25 mm resulted in asymmetric, unstable leading vortex along with the transmitral jet. Based on their *in vitro* results, they hypothesized that flow interaction with the ventricular wall can significantly influence the stability, dynamic and the shape of the leading vortex that develops along with transmitral jet.

3.1.4 Transmitral Vortex Formation Time Index: *A Parameter to Couple Diastole and Systole*

Diastole and systole are two functional pumping modes of the LV. Although they are functionally independent and can be individually dysfunctional, they should be related to each other based on the conservation of mass principle. In that sense, the blood volume that enters the LV during diastole should be equal to the volume of blood that leaves the LV during systole. In a cardiac cycle, blood enters to the LV during early diastole and atrial contraction phases, described by E- and A-waves of PW Doppler echocardiography. Vortex formation time (VFT) can be computed from the governing equations for transmitral flow and ejection fraction (EF) as (Gharib et al. 2006):

$$VFT = \frac{4(1-\beta)}{\pi}\alpha^3 \times EF \tag{3.2}$$

where β is the fraction of stroke volume contributed from the atrial component of LV filling obtainable as:

$$\beta = \frac{V_A}{EDV} = \frac{VTI_A \times \frac{\pi}{4} D_E^2}{EDV} \tag{3.3}$$

where V_A is the blood volume that enters the LV during atrial contraction, *EDV* is the left ventricular end-diastolic volume, VTI_A is the velocity-time integral of A-wave, and D_E is the effective diameter of mitral geometric orifice area (*GOA*; Fig. 3.3) (Kheradvar and Gharib 2009):

$$D_E = 2\sqrt{\frac{GOA}{\pi}} \tag{3.4}$$

The parameter α^3 is a non-dimensional volumetric parameter for the LV, obtained by dividing the EDV by cubic power of D_E.

$$\alpha = \left(\frac{EDV}{D_E^3}\right)^{\frac{1}{3}} \tag{3.5}$$

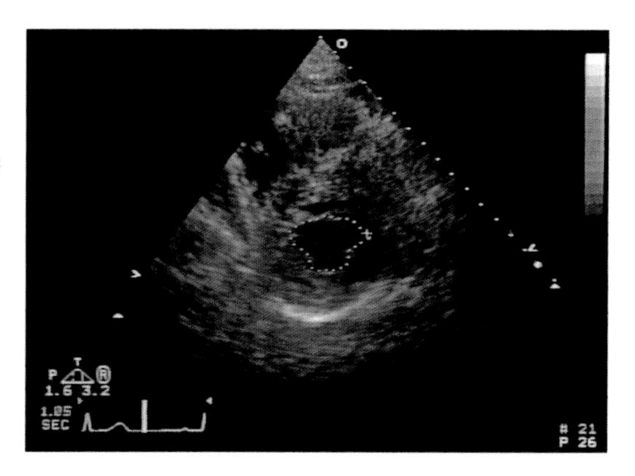

Fig. 3.3 The Geometric Orifice Area (GOA) of mitral valve. To acquire the effective diameter, an equivalent circular shape area is considered. Then the diameter of the equivalent circle is computed as the effective diameter of GOA (D_E). Modified from Kheradvar et al. (2011)

3.1.5 Mitral Annulus Recoil

Dynamic nature of mitral annulus motion has been previously verified in human (Jimenez et al. 2003; Ryan et al. 2007) and in animal models (Karlsson et al. 1998; Salgo et al. 2002). Carlhall et al. (2004) found that the excursion of the mitral annulus significantly accounts for the total ventricular filling and emptying in human since the annulus plays a sphincter-like role facilitating ventricular filling and valve closure during diastole and systole, respectively. Through early diastole, the mitral annulus movement exhibits rapid recoil back toward the LA that increases the net velocity of the mitral inflow as much as 20% (Keren et al. 1988). Mitral annulus moves toward the LA during diastole and toward the ventricular apex during systole. The motion of the mitral annulus during the rapid filling phase represents the recoil force applied to the base of the heart. Several factors such as transmitral flow, ventricular compliance and papillary muscle contraction influence the dynamics of mitral valve annulus (Kheradvar and Gharib 2007, 2009). Altered ventricular geometry, impaired relaxation and/or viscous deformations can also influence the mitral annulus recoil by affecting the transmitral pressure gradient. Kheradvar and Gharib showed *in vitro* that the dynamic recoil of mitral annulus is strongly correlated to the process of transmitral vortex formation such that the peak annulus recoil occurs once the transmitral vortex is pinched off from the jet (Kheradvar and Gharib 2007; Kheradvar et al. 2007). The geometry of leaflets would also affect the recoil force. Additionally, Kheradvar and Gharib showed that the presence of leaflets would dissipate the annulus recoil. They hypothesized that the damping effect can be explained by the birth of three-dimensionality in transmitral jet resulted from changes in orifice eccentricity due to the leaflets. Finite changes in orifice eccentricity generate a fully three-dimensional vortex wake that evolves quite differently from the ones developed under axisymmetric conditions (Bolzon et al. 2003). Therefore, the diastolic vortex deviates from quasi-axisymmetric geometry due to the leaflet interaction, and in fact, deforms progressively because of the self-induced motions. Accordingly, there are viscous reconnections of the wake vortex lines adjacent to the ventricular wall with the boundary layer vortex lines during the early phase of wake dissipation (Bolzon et al. 2003).

3.1.6 Grading Diastolic Dysfunction

It has been shown that the traditional echocardiographic criteria for diagnosis and grading cardiac dysfunction may not be sufficiently reliable (Petrie et al. 2004; Ogunyankin et al. 2006) and several measurements are needed to characterize the dysfunction. This is particularly true with respect to diastolic dysfunction (Dokainish et al. 2008; Jenkins et al. 2005; Andrew 2003). The key challenge in interpretation of Doppler mitral inflow patterns is in distinguishing a normal from pseudonormal pattern (Xie and Smith 2002). Despite the importance of early diagnosis for diastolic dysfunction, prompt characterization of dysfunction still remains unclear. The method of choice for LV assessment -cardiac catheterization- is invasive and inappropriate for screening purposes. Conventional non-invasive Doppler measurements have low sensitivity, specificity and predictive accuracy (Zile and Brutsaert 2002; Khouri et al. 2004; Oh et al. 1997). Additionally, the majority of the indices such as mitral valve inflow and pulmonary venous flow are load-dependent and obtaining satisfactory measurements of pulmonary venous flow (Khouri et al. 2004) or mitral inflow by transthoracic echocardiography (TTE) is difficult at peak Valsalva maneuver (Ommen et al. 2000). Doppler tissue imaging (DTI) measures the early diastolic myocardial velocity at the lateral corner of the mitral annulus (e'), which is shown to be a more reliable marker for LV relaxation while appears to be quite load independent (Sohn et al. 1997). In spite of this, major limitation of this index is the extrapolation of regional motion to predict global diastolic function. In addition, several studies have questioned its load independence in situations affected by acute changes in preload (Dincer et al. 2002; Firstenberg et al. 2001). Recent results also indicate that despite e' is significantly different in normal controls versus diastolic dysfunction subjects, the index cannot distinguish between varying degrees of dysfunction (Kheradvar et al. 2011). In other words, the e' index may not be sensitive enough to distinguish between grades of diastolic dysfunction. This observation may provide evidence that when mitral annuls recoil is impaired, variation in transmitral vortex formation determines the pattern of clinical presentation.

To accurately determine the grade of diastolic dysfunction, we currently need to measure several indices as stated above. However, each individual index obtained from Doppler measurements is restricted only to a certain aspect of diastolic function, and cannot be independently regarded as a global representative for the left heart diastolic function. The combination of various Doppler indexes, and sometimes cardiac catheterization, is required to accurately interpret the diastolic function whereas every single index has its own inaccuracies and limitations (Najos-Valencia et al. 2002; Whalley et al. 2005), which may result in misleading interpretations. It is suggested that the transmitral vortex formation time (VFT) is a single index that may contribute to distinguishing different grades of diastolic dysfunction while can be obtained easily with low cost. This index and other measures based on the transmitral vortex may potentially be used for determination of the disease prognosis and may provide further guidance toward evaluation of the therapeutic strategies in future.

3.1.7 Outcome Planning for Diastolic Dysfunction

Great heterogeneity exists for results in prognosis of diastolic dysfunction (Galderisi 2005) which clearly emphasize on the importance of the evaluation of LV diastolic properties in clinical research. Two important studies revealed the prognostic value for grading diastolic dysfunction (Schillaci et al. 2002; Bella et al. 2002). The PIUMA study (Schillaci et al. 2002), evidenced that the pattern of abnormal relaxation increases the risk of cardiovascular events (odds ratio 1:57, 95% CI 1.1–2.18, $p < 0.01$) in a population of 1839 hypertensive patients during a 11 years follow-up. The Strong Heart Study (Bella et al. 2002), during a 3-year follow-up on a population of 3,008 American Indians, showed that an abnormal relaxation pattern was associated to a two-fold increase of mortality risk while pseudonormal/restrictive patterns were associated to a three-fold increase of cardiac mortality. This result is consistent with the findings of the Framingham Heart Study, where a “U” relation between transmitral A-wave and the risk of atrial fibrillation is detectable, and the arrhythmia appears to be independently associated with both abnormal relaxation and pseudonormal/restrictive patterns (Bella et al. 2002). These two studies, particularly the Strong Heart Study, are very consistent with the pathophysiologic standpoint of the Mayo Clinic investigators, who created an ingenious classification of Doppler-derived Diastolic dysfunction few years ago (Nishimura and Tajik 1997). In this classification, the pattern of abnormal relaxation (grade I of DD) and both reversible and irreversible restrictive patterns (grade III and IV, respectively) were at opposite sides in the clinical progression towards the end stages of heart failure while the pseudonormal pattern has an intermediate, but clinically crucial position. In consideration of these findings and combining the value of the prognostic studies, defining novel indices that quantify transmitral flow, such as vortex formation time index that can differentiate the stages of diastolic dysfunction, would be extremely useful in clinical follow-up and potentially in assessment of response to treatment in future. These indices may provide an important step forward in the evaluation of diastolic function and determination of the disease prognosis.

3.2 Aortic Valve and Sinuses of Valsalva

The anatomy and function of the aortic valve have inspired scientific interests for the past 600 years beginning with Leonardo da Vinci who studied aortic valve and the role of the sinuses of Valsalva in terms of cusp closure motion (Robicsek 1991). Leonardo's study of the aortic valve, as illustrated in his drawings was extraordinary (da Vinci 1513). Leonardo's work on the shape of the valve cusps suggested that the outflow through the triangular orifice made a 3D symmetric vortex as illustrated in his sketches on heart valves. He hypothesized that the vortex causes the valve to close, preparing for the next pulse of blood. Based on his sketches, Leonardo also found that the central, most powerful part of the blood jet ejected through the aortic valve produces the necessary pressure wave through the aorta and the rest of the arteries as he recognized that by palpable pulse. These long-lasting concepts are reviewed here on the basis of the current knowledge and modern technology.

3.2.1 *Functional Anatomy*

Aortic valve is situated in the left ventricular outflow tract where the aorta begins. It consists of three lunar-shaped pocket-like flaps of tissue, referred to as cusps. The leaflets of the aortic valve are attached partially to the muscular walls of the LV. In contrary to mitral valve, the aortic valve leaflets are not supported by ring-shape structure but rather in a crown-like fashion (Anderson 2000). During diastole while the LV is being filled with blood through mitral valve, the aortic valve remains closed. Once it is closed, the cusps are coapting, aligned and separate the LV from the aorta.

The Valsalva sinuses are dilatations between the aortic wall and each of the semilunar cusps of the aortic valve. Generally, there are three aortic sinuses, the left, the right and the posterior. The left aortic sinus gives rise to the left coronary artery, and the right aortic sinus gives rise to the right coronary artery. Usually, no vessels arise from the posterior aortic sinus, which is known as the non-coronary sinus.

3.2.2 *Vortex Formation in Aortic Sinus*

The aortic blood flow during systole is considered an unsteady mainstream flow along with vortices that spin within the sinuses of Valsalva (Peacock 1990). These vortices have been of interest since Leonardo da Vinci first described them in the fifteenth century (da Vinci 1513). As the aortic flow decelerates during late diastole, no variation in radial velocity can be observed, except the thin boundary layer that travels alongside the aortic wall, distal to the aortic sinuses. Within the outer laminas of the mainstream aortic flow, the velocity changes from zero at the wall to its peak midstream magnitude. As this layer reaches the tip of the fully open aortic valve leaflets, it separates from the wall, and continues downstream as a free shear layer. Once this separated shear layer is cut off by the sinus ridge, a portion of the fluid begins to curl back toward the ventricle, which results in a spinning vortex formed within the Valsalva sinus cavities (Peacock 1990).

The very first attempt to quantitatively understand the aortic sinus vortex was made by Bellhouse and Talbot who assumed that the flow in each sinus can be described by half of a Hill spherical vortex (Bellhouse and Talbot 1969). Their model suggests that closure of the valve is being started during the deceleration phase of the aortic flow, and the valve is almost being closed at the end of systole. Alternatively, (Peskin and Wolfe 1978) described a substitute theory based on dynamics of a point vortex in a two-dimensional irrotational flow. Their approach takes the motion of the vortex into account, and defines the mechanisms through which the aortic sinus traps a vortex in terms of stability of the equilibrium position of the vortex. The *point vortex theory of the aortic sinus* is derived from a conformal mapping of the fluid domain that results from the merger of the upper half-plane (aorta) and an overlapping circular disc (sinus) (Peskin 1982).

This analysis concludes that the sinus vortex has an equilibrium position along the midline, which is uniquely determined by assuming that a streamline separating from the upstream border of the sinus reattaches at the downstream border. Thus, the equilibrium vortex strength is proportional to the free-stream velocity while its equilibrium position is independent of the free-stream velocity. Considering the vortex trajectories in the neighborhood of the equilibrium point as they form closed loops, the stability result is due to entrapment of the vortex in the sinus (Peskin 1982). In pulsatile flow through aorta, the viscous forces along the wall and the leaflet generates a strong vortex trapped in the sinus cavity. The presence of such a strong vortex potentially generates a pressure distribution in the plane of the leaflet that can close the valve when the aortic flow is decelerating (Peskin and Wolfe 1978; Peskin 1982; Van Steenhoven and Van Dongen 1979).

3.3 Vortex Formation in the Right Heart

Currently, only few studies have been performed to characterize right ventricular (RV) fluid dynamics. RV function has been shown to be a major determinant of clinical outcome and should be considered during clinical management and treatment of cardiac dysfunction (Graham et al. 2000; Burgess et al. 2002; D'Alonzo et al. 1991; Mehta et al. 2001; Zehender et al. 1994; de Groote et al. 1998). The common causes of RV dysfunction are broadly divided into: (a) intrinsic RV failure in the absence of pulmonary hypertension that usually occurs due to RV infarction; (b) RV failure secondary to increased RV afterload; and (c) RV failure because of volume overload. Evaluation of RV is extremely challenging due to the anatomic and functional complexities of this chamber. Principally, RV function cannot reliably be evaluated by conventional 2D echocardiography techniques due to its asymmetrical lunar shape, narrow acoustic window and geometrical assumptions for calculations of volume (Silverman and Hudson 1983; Levine et al. 1984). Furthermore, clinical assessment of RV dysfunction is particularly difficult as most patients do not show clinical symptoms of systemic venous congestion in the early stages of disease (Tayyareci et al. 2008).

The RV was long considered a relatively passive chamber for blood flow between the systemic and pulmonary circulations until recent studies emphasized on its significance in maintaining hemodynamic stability and overall cardiac function (Mehta et al. 2001; Kevin and Barnard 2007). It is worth to mention that the RV ejection fraction is directly influenced by the total pulmonary resistance, which is critically affected by left ventricular end-diastolic pressure and pulmonary vascular resistance (Bleasdale and Frenneaux 2002; La Vecchia et al. 2001; Nakamura et al. 1994). Despite the improvements in noninvasive intraventricular flow visualization approaches, several aspects of the RV fluid dynamics phenomena are still unclear. The key reason for lack of quantitative information concerning cardiac flow is the lack of technologies that can reliably map 3D spatial and temporal details of RV flow while being feasible in routine clinical use.

The flow inside the RV is expected to form a complex three-dimensional (3D) arrangement involving vortex formation from the tricuspid jet as well as rotation about longitudinally-oriented axes. The complexity of the RV flow field was first demonstrated though numerical simulations by (Peskin and McQueen 1996). Despite the reduced accuracy and resolution of the numerical solution, the distorted nature of the large scale circulation was appreciable. The only other data on characterizing diastolic RV vortex in medical literature is based on the study performed by (Pasipoularides et al. 2003) on dogs using computational fluid dynamics simulations on a CRAY T-90 supercomputer. Their results also corroborate on formation of 3D large-scale vortices in a highly complex flow field.

More recent computational analyses have been carried out on RV hemodynamics with the aid of modern cardiac imaging tools (Mangual et al. 2011). The three-dimensional RV endocardial geometry was reconstructed from echocardiography (4D RV-Function software, TomTec Imaging Systems GmbH) and used as the boundary conditions (Fig. 3.4) to solve for the Navier-Stokes and continuity equations within the RV cavity. These were solved using a fast spectral – finite difference method (Fadlun et al. 2000) and the resulting flow was analyzed in terms of the 3D vortex dynamics.

Figure 3.5 illustrates vortex formation in a sagittal cut of the RV during one cardiac cycle; red and blue correspond to clockwise and counterclockwise rotations, respectively. During early diastole, a vortex ring is formed past the tricuspid valve, and at late diastole it breaks down into an irregular pattern of smaller vortex elements. During systole, the flow becomes apparently more regular while moving through the outflow tract. A 3D representation is shown in Fig. 3.6 for early diastole and late systole, with sagittal and coronal cross sectional cuts of the vortical pattern in the RV. The superimposed vortex structure is identified by the λ_2 method as described in Sect. 2.5. In early diastole, a quasi-perfect vortex ring is observed in the center of the

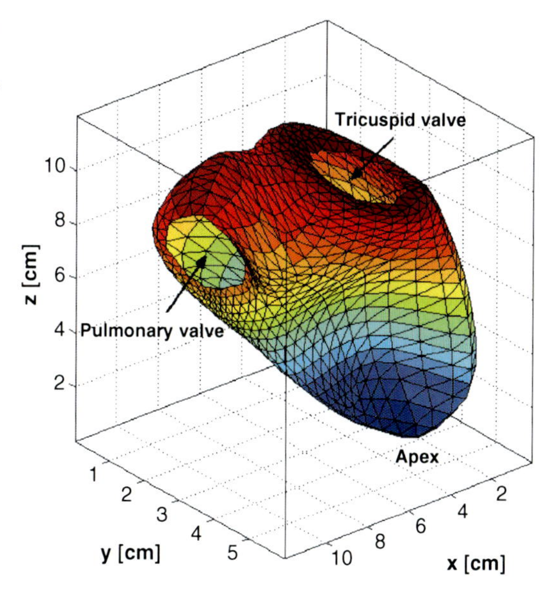

Fig. 3.4 3D RV geometry extracted from echocardiography imaging used in the computational simulation

RV, past the tricuspid valve (gray isosurface). As the filling progresses, vortex ring interacts with the interventricular septum, and dissipates on that side. Therefore, the vortex undergoes instability and flow within the RV becomes highly disturbed mainly due to the complex crescent-shape geometry of the RV. During systole, the vorticity inside the RV is mainly due to the breakdown of the leading transtricuspid vortex that rearranges into a streamwise vortex filament toward the outflow tract. This highly vortical flow may promote proper mixing of the deoxygenated blood in the RV, thus prevents blood stagnation. Further studies are required to clearly understand the complex flow in the right heart.

3.4 Vortex Formation in the Embryonic Heart

The developing mammalian heart grows from a simple valveless pump into a four-chambered heart over a wide range of scales. During its growth, the heart continues to function and drive blood flow to ensure proper development of the vascular system. The flow within the developing heart is thought to act as an epigenetic signal for chamber and valve morphogenesis. It is speculated that the variation in cardiac morphology correspond to fluid dynamic changes related to scaling effects. Since congenital heart defects affect 1% of all live births, they are responsible for a vast majority of prenatal losses (Gruber and Epstein 2004). Insight into the fluid dynamics of the embryonic heart and its contribution to development, is critical towards understanding these malformations and improving prenatal care.

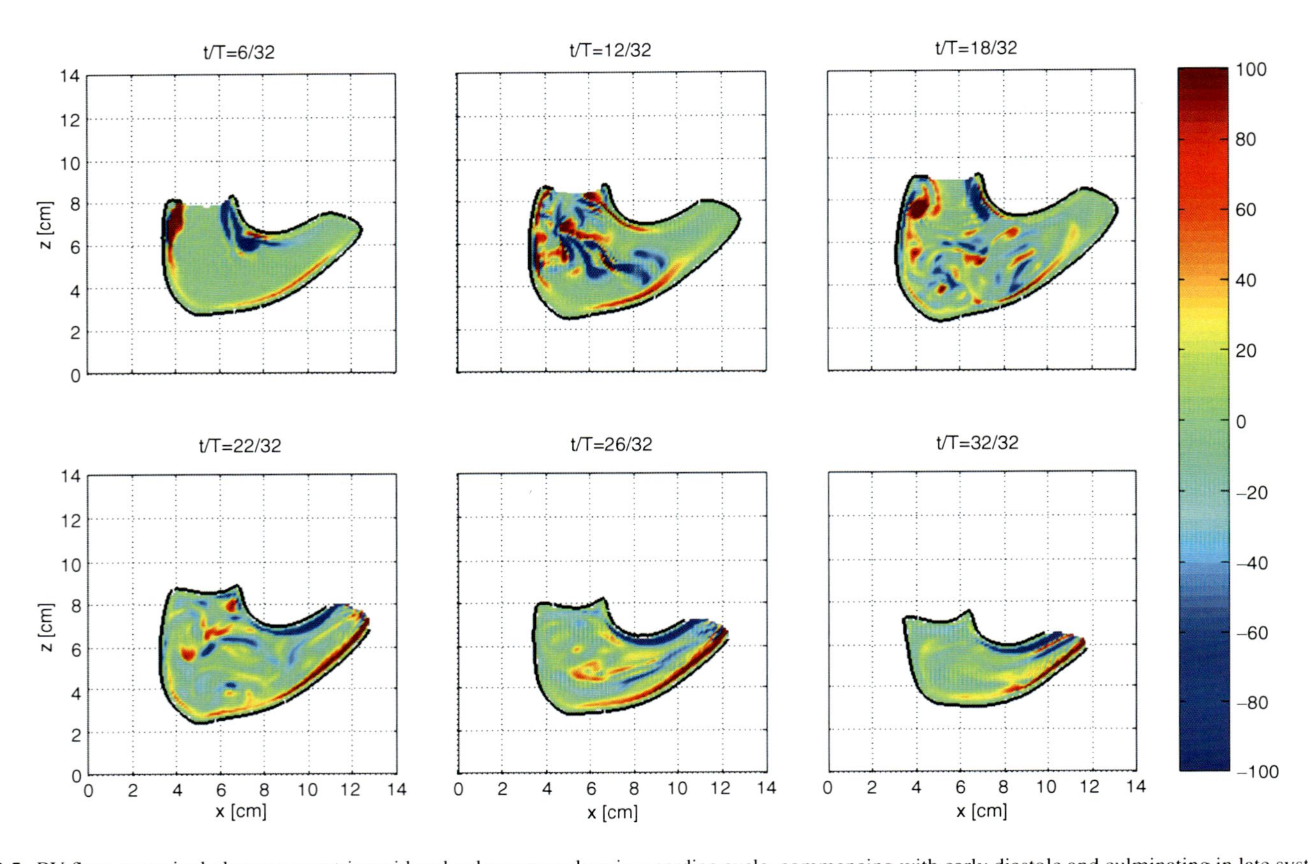

Fig. 3.5 RV flow on sagittal planes across tricuspid and pulmonary valves in a cardiac cycle, commencing with early diastole and culminating in late systole. T is the normalized time corresponding to a cardiac cycle. Frames show the normal vorticity component distribution; vorticity values (s^{-1}) are shown in the *color bar*

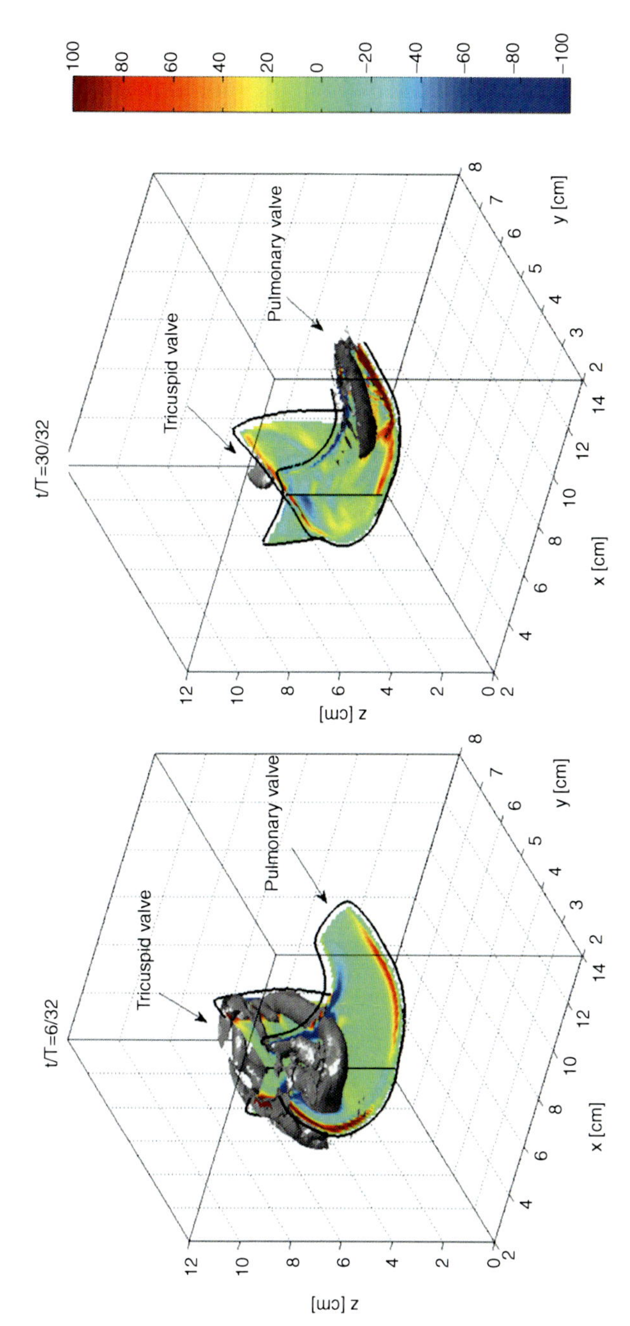

Fig. 3.6 Transversal planes along the tricuspid and pulmonary valve and perpendicular plane along the tricuspid valve of the RV with normal vorticity component fields and superimposed 3D vortex structure, λ_2. Plots correspond to diastole (t/T = 6/32) and systole (t/T = 30/32). T is the time corresponding to one heartbeat. Vorticity values, s^{-1}, are shown in the color bar. 3D vortex structure visualized by isosurface of λ_2 field, value $\lambda_2 = -900\ s^{-2}$

Table 3.1 Reynolds and Womersley numbers for vertebrate embryonic hearts and aortas (shaded in gray) at several stages of development given in hours post fertilization (h.p.f.), days post fertilization (d.p.f.), days post conception (d.p.c.), and Hamburger & Hamilton stages (HH)

Species	Flow rate (mL/s)	Diameter (mm)	Frequency (Hz)	*Wo*	*Re*	Reference
Zebrafish, 26 h.p.f.	1	0.05	2.3	0.111	0.017	Forouhar et al. (2006)
Zebrafish, 4.5 d.p.f.	10	0.1	2	0.207	0.342	Hove et al. 2003
Chicken, HH15	26	0.2	2	0.414	1.777	Vennemann et al. (2006)
Mouse, 8.5 d.p.c.	3	0.075	2.8	0.184	0.077	Jones et al. (2004)
Mouse, 9.5 d.p.c.	4	0.125	2.1	0.265	0.171	Jones et al. (2004)
Mouse, 10.5 d.p.c.	4	0.15	2.4	0.340	0.205	Jones et al. (2004)
Chicken, Stage 18	170	0.083	2	0.172	4.82	Wang et al. (2009)
Chicken, Stage 24	250	0.14	2	0.290	11.96	Wang et al. (2009)
Mouse, 11.5 d.p.c.	127	0.33	3.78	0.940	14.32	Phoon et al. (2002)
Mouse, 12.5 d.p.c.	158	0.36	4.07	1.064	19.43	Phoon et al. (2002)
Mouse, 13.5 d.p.c.	173	0.35	4.40	1.076	20.69	Phoon et al. (2002)
Mouse, 14.5 d.p.c.	226	0.34	4.35	1.039	26.25	Phoon et al. (2002)

The dynamic viscosity and density of the blood was set to 0.003 N s/m^2 and 1,025 kg/m^3, respectively. The characteristic velocities and length scales were set to the peak flow rate and the maximum diameter of the heart or aorta. Note that the calculation of the dimensionless numbers is sensitive to the choice of the characteristic length, velocity, viscosity, and density. These calculations may be different than those reported in the references

Several dimensionless parameters obtained by non-dimensionalizing the Navier-Stokes equations are currently in use for understanding fluid dynamics scaling effects. Two of the most commonly used dimensionless numbers are the Reynolds number (*Re*) and the Womersley number (*Wo*), as previously introduced in Sect. 1.6.

The *Re* can be considered as the ratio of inertial forces to viscous forces acting in the fluid, and the *Wo* characterizes the significance of unsteady effects. Macroscopic flows, such as those in the aorta, are typically inertia dominated ($Re >> 1$) with significant pulsatile effects ($Wo >> 1$). Microscopic flows ($Re << 1$), such as those observed during the early embryonic stages of heart development, are typically viscous dominated and their unsteadiness is not significant ($Wo << 1$). A summary for *Re* and *Wo* of the heart and aorta in different species during development is shown in Table 3.1.

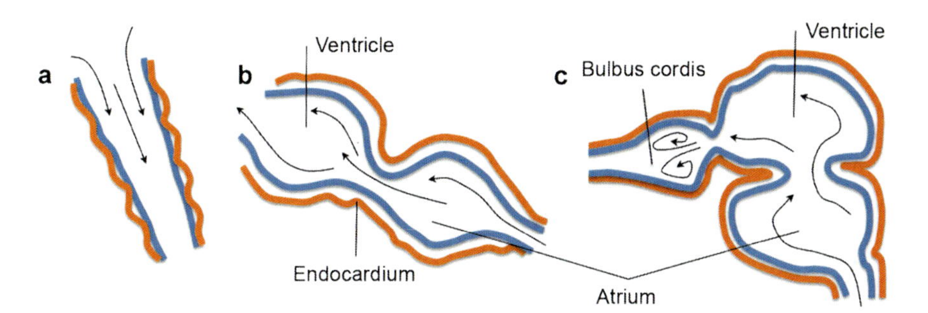

Fig. 3.7 Schematic diagram of the zebrafish embryonic heart during development. (**a**) At 24 h.p.f., the heart is a simple, valveless linear tube and the flow of blood is unidirectional. (**b**) The early atrium and ventricle are evident at about 30 h.p.f., and the blood flow remains unidirectional as it traverses through the chambers and over the developing cardiac cushions. (**c**) Cardiac looping has completed, and the chambers have formed in the zebrafish heart at 4.5 d.p.f. Vortices appear behind the atrio-ventricular constriction during ventricular filling, and a vortex pair forms in the bulbus arteriosus during ventricular systole

One important change in the intracardial blood flow pattern during development is the formation of vortices in the chambers and the aorta. Cardiac endothelial cells can detect direction and magnitude of the flow, where such shear signals are thought to trigger the biochemical pathways that activate some of the genes required for morphological development (Jones et al. 2004). The presence or absence of vortices in the chamber and aorta would alter the magnitude and direction of shear stresses, and may also affect the mixing patterns of the blood. In particular, the intrachamber vortices are important in assisting the cardiac cushion closure during cardiac looping and chamber morphogenesis. Vortex formation in the aortic sinus is also significant for the proper development and function of the aortic valves.

In all vertebrate animals, the heart initially forms as a linear valveless tube that drives blood via peristalsis or valveless suction pumping (Santhanakrishnan and Miller 2011). Forouhar et al. (2006) used particle image velocimetry to spatially reconstruct the flow fields in the zebra fish tubular heart 36 hour post fertilization (h.p.f.). They found that the blood flows smoothly through the tube (Fig. 3.7a) and did not observe the presence of vortices. Vennemann et al. (2006) also used particle image velocimetry to reconstruct the flow fields in HH15 (Hamburger and Hamilton 1951) chick tubular hearts and reported similar flow patterns. They synchronized their measurements to different portions of the cardiac cycle and did not report formation of vortices. The initial vortex formation has been observed towards the end of cardiac looping and chamber morphogenesis as the Reynolds Number increases and inertia becomes relevant to the flow. Hove et al. (2003) reported the formation of vortices within the atrium, ventricle, and bulbus arteriosus 4.5 days post fertilization (d.p.f) in wild-type zebra fish hearts (Fig. 3.7b, c). The strongest vortices were observed in the bulbus arteriosus as the blood was pumped from the ventricle. Chamber vortices were weaker and appeared to be on the same scale of the red blood cells. These vortices may be generated or strengthened by tumbling and rotation of the red blood cells in the flow.

Computational fluid dynamics provides insight into the complicated flow structures within the developing heart chambers and aorta. DeGroff et al. (2003) constructed a three-dimensional computational model of the surface of the human heart at stages 10 and 11 using a sequence of two-dimensional cross-sectional images. The computational model was constructed such that the pulsatile flow was driven through a rigid heart. Coherent vortices were not observed in their numerical simulation but rather streaming was present in the heart tube. In other words, particles released on one side of the lumen did not cross over or mix with particles released from the opposite side. Liu et al. (2007) used a three-dimensional finite element model of a HH21 chick embryonic heart to quantify hemodynamic forces on the outflow tract. Their computational model included wall flexibility but did not include cardiac cushions. Pairs of counter-rotating vortices were observed during the ejection phase near the inner curvature of the outflow tract. Maximum velocities were observed in regions of constrictions, corresponding to a maximum *Re* of 6.9.

To investigate the effects of scale and morphology on vortex formation, Santhanakrishnan et al. (2009) used simple physical and mathematical models of the heart when the atria, ventricle, and cardiac cushions begin to form as depicted in Fig. 3.8. They found that the conditions required for vortex formation are significantly affected by *Re* and are highly sensitive to chamber depth and cushion height. Flow over models of the ventricle at three *Re* are shown in Fig. 3.9. At scales appropriate to the initial formation of the heart tube ($Re \approx 0.05$), flow is unidirectional and moves smoothly over the cushions, and through the chambers. This can be visualized as the red lines of dye, which were initially drawn vertically in the chambers, deform while the fluid moves through the models (Fig. 3.9a). The next panels show the patterns of flow for $Re \approx 40$ in presence (c) or absence (b) of cardiac cushions. The presence of cushions triggers the formation of the chamber vortex, and the direction of flow reverses along the chamber walls. Generally, presence of vortices is particularly sensitive to changes in chamber depth and cushion height for *Re* in this range. Pulsatile effects on transition to vortical flow were not yet investigated, but it is likely that the presence or absence of vortices in this intermediate range could be triggered through unsteadiness in the fluid since *Wo* is on the order of one or more. Finally, presence of chamber vortices is shown by injecting dye upstream of the model ventricle for $Re \approx 1{,}000$, corresponding to the scale of the adult heart (Fig. 3.9d). Chamber vortices are always present for realistic chamber depths, in absence or presence of the valves. In summary, large scale structure of the blood flow is critically sensitive to small changes in scale and morphology.

Intracardiac forces generated by blood flow are essential for proper cardiovascular development and are especially critical for proper formation of the valves. The specific details of the magnitude and direction of the blood flow during development, particularly at the level of endothelial shear sensing, have not been well resolved. Accurate descriptions of normal intracardiac flow profiles and the formation of chamber and aortic vortices during each stage of development are critical to the diagnosis and correction of congenital heart defects. *In utero* surgical interventions have been used to correct severe aortic stenosis, improving ventricular function and preventing the development of hypoplastic left heart syndrome (HLHS) (Wilkins-Haug et al. 2005; Selamet

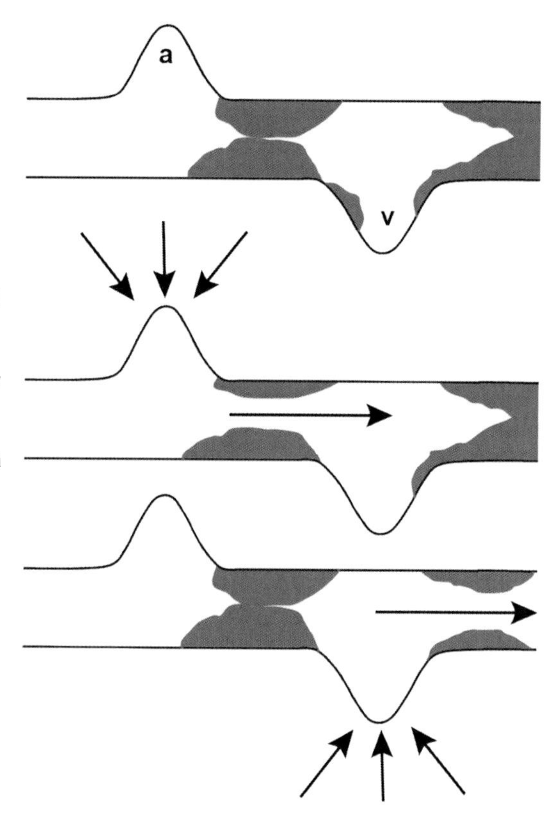

Fig. 3.8 Schematic diagram of the vertebrate embryonic heart during chamber formation, drawn after Moorman et al. (2004). The developing cardiac cushions are shown in *gray*, and *a* and *v* indicate the locations of the atria and ventricle respectively. Alternating contractions of the atrium and ventricle drive the blood within the developing heart. The fluid flow and contraction patterns are presented with increasing time, looking from *top to bottom*. *External arrows* show the direction of chamber contractions, and *arrows within* the heart denote the blood flow direction

Tierney et al. 2007). *In utero* echocardiography has been used to detect abnormalities related to structural heart diseases such as univentricular heart (UVH), ventricular septal defect (VSD), and HLHS from 16 weeks onward (Boldt et al. 2002). Improved understanding of normal and pathological blood flow patterns during development may substantially improve *in utero* diagnostics.

3.5 Linking Cardiac Muscle Function to Vortex Formation

The form and function of the heart create time-varying and spatially-complex pulsatile flow patterns that are optimal for minimizing energy dissipation. One of the most intriguing features of the intracardiac flow patterns is the formation of intracavitary vortices. Investigators have suggested that the asymmetric geometry of the LV and the orientation of its valves may favor vortex formation and automatic redirection of blood from the LV inflow to the outflow. For example, Pedrizzetti and Domenichini in a numerical model, showed that asymmetric circulation in the LV

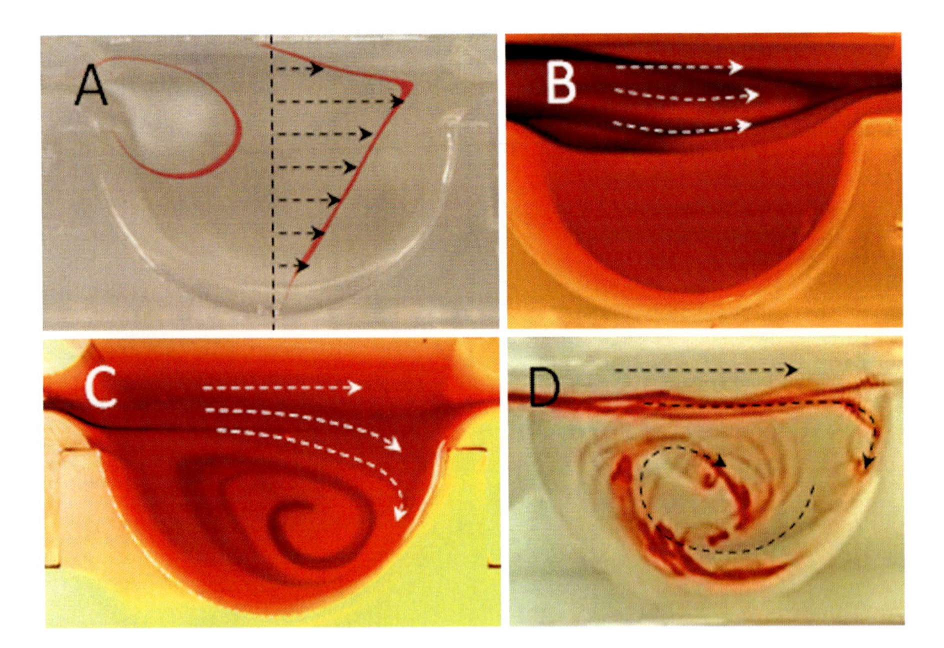

Fig. 3.9 Flow visualization within physical models of the early ventricle showing the effect of varying *Re* and cushion morphology on vortex formation. Flow direction is from left to right. (**a**) At $Re \approx 0.05$, the flow is unidirectional as evidenced by the deformation of lines of dye initially drawn vertically. (**b, c**) The presence of chamber vortices can be triggered by the increased height of the cardiac cushions at $Re \approx 40$, as shown by *color changes* generated upstream using the pH indicator bromothymol blue. (**d**) For realistic chamber depths, vortices are always observed at the high *Re* characteristic of the adult heart ($Re \approx 1{,}000$). The presence of chamber vortices results in a change in the magnitude and direction of shear at the chamber wall

largely depends on the physiologic eccentricity of the mitral orifice (Pedrizzetti and Domenichini 2005). If the mitral orifice was oriented more centrally, energy dissipation would increase since the flow pattern at the end-diastole may not be favorable for ejection, and fluid elements would be subjected to more strain for being redirected toward the aorta. In contrast, when the mitral orifice is highly eccentric, the transmitral flow collides with the ventricular wall, enhancing the dissipative interaction with the boundary layers, which promotes natural swirling of blood flow inside the LV and facilitate the ejection of blood.

Recent techniques that measure myocardial motion by speckle tracking in grayscale images have overcome the angle dependence of Doppler Tissue Imaging (DTI) strain, allowing for noninvasive measuring of the parameters that define cardiac muscle mechanics. DTI, DTI strain, and speckle tracking may provide unique information that decipher the deformation sequence of myofibers oriented in complex form within the LV wall. The LV flow sequence (Fig. 3.10) is assumed to be linked to the temporal sequence of deformation seen in the LV (Sengupta et al. 2005, 2006).

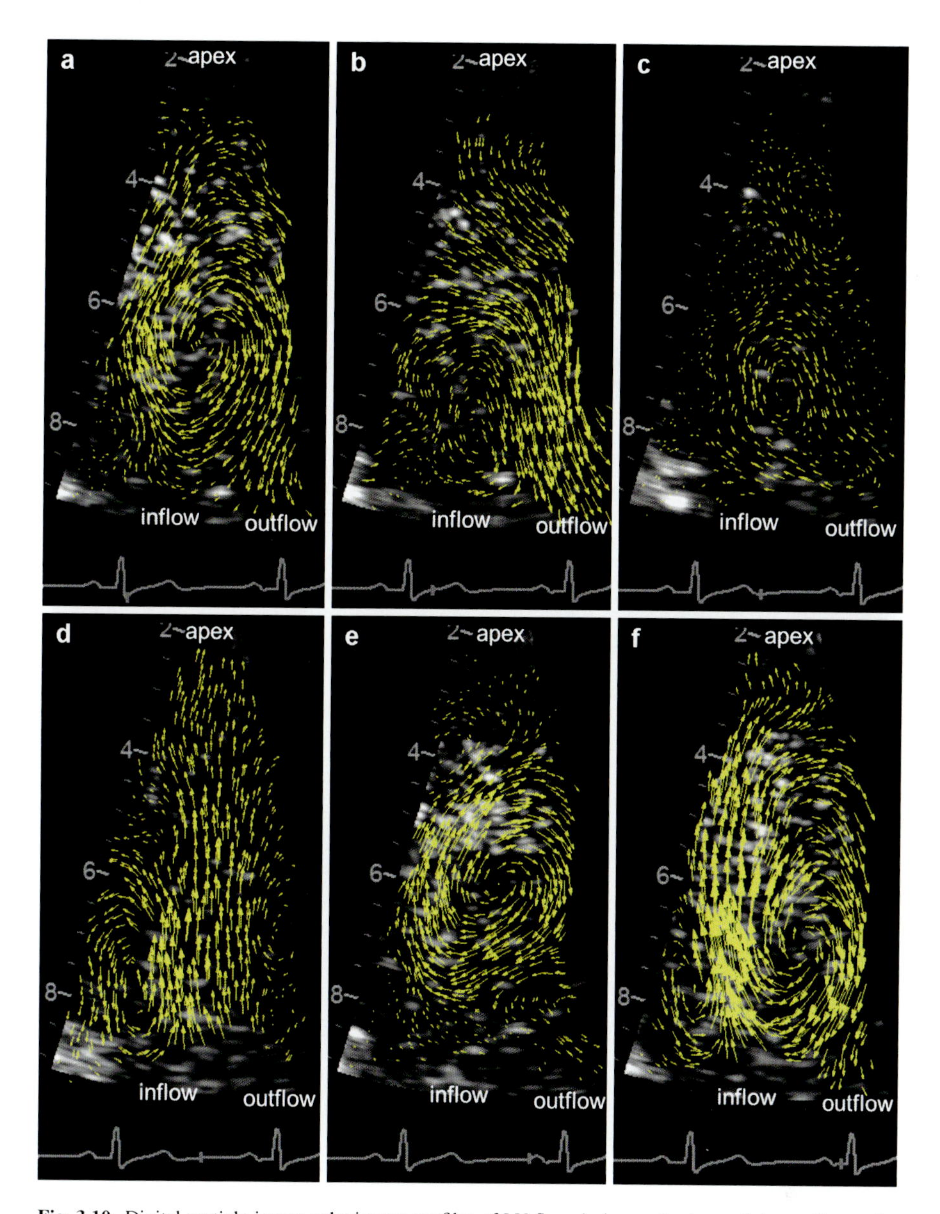

Fig. 3.10 Digital particle image velocimetry profiles of LV flow during each phase of the cardiac cycle. Under specific conditions of dilution and administration to blood circulation, the echo contrast particles (*microbubbles*) can be tracked for calculating vectors and trajectories of flow within a 2-dimensional ultrasound scan plane. The velocity vector fields are shown during isovolumic contraction (**a**), ejection (**b**), isovolumic relaxation (**c**), early diastole (**d**), diastasis (**e**), and late diastole (**f**). Note the apex-to-base redirection of blood flow during isovolumic contraction with formation of a vortex across the inflow-outflow region and the base-to-apex reversal of blood flow during isovolumic relaxation.

3.5.1 Preejection

A rapid apico-basal spread of electrical activation within the subendocardium initiates the contraction sequence. The onset of subendocardial shortening coincides hemodynamically with an early rapid build-up of intraventricular pressure during the isovolumic contraction period. Subepicardial deformation occurs later, temporally coinciding with the onset of systolic ejection. The temporal lag between the onsets of subendocardial and subepicardial contraction activities correlates with the duration of the isovolumic contraction (IVC). Shortening and lengthening of the myocardial wall also result in rotary movements mainly due to the intrinsic spiral geometry of the myofibers. During IVC phase, predominant shortening of the subendocardial fibers and stretching the subepicardial fibers result in a brief clockwise rotation of the apex (right-handed helix rotation directed along subendocardial fiber direction).

During the pre-ejection phase -prior to closure of the mitral valve- blood flow begins to accelerate towards the base, consistent with the apex-to-base direction of electromechanical activation. This stream of blood accentuates the trailing vortex formed across the anterior edge of the closing anterior mitral leaflet. The rearrangement and redirection of blood flow during the preejection phase temporally coincides with the biphasic movement of the LV wall observed on DTI. Contraction in one direction (right-handed helix) and stretching in the orthogonal direction (left-handed helix) assist transfer of the intracavitary blood toward the LV outflow. The biphasic deformations satisfy isovolumic mechanics; shortening in one direction is accompanied with stretching in the other direction. Stretching of the subepicardial myofibers initiates an intrinsic length-sensing mechanism (stretch activation), which allows muscle to adjust the force of subsequent shortening during ejection (Campbell and Chandra 2006).

3.5.2 Ejection

During ejection phase, the ventricular cavity shrinks in all directions due to contraction of both subendocardial and subepicardial fibers. However, shortening strains in the apical segments exceed those in the basal segments such that the wave of shortening moves in the axial direction from the apex toward the base (Sengupta et al. 2006; Prinzen et al. 1992; Buckberg et al. 2006). Also transmural shortening is accompanied with rotation, where the direction of rotation is governed by the subepicardial fibers owing to their longer arm of movement (Taber et al. 1996) and their intrinsic contractile properties (Davis et al. 2001). Shortening of subepicardial fibers results in counterclockwise rotation of the LV apex and clockwise rotation of the LV base. The net result is twist or torsion that results in a wringing movement of the LV. This results in axial propulsion of blood from the LV cavity during ejection.

3.5.3 Isovolumic Relaxation

Reversal of the shortening-lengthening relationship defines the physiological onset of relaxation. A physiological postsystolic shortening of the LV myocardium creates active apex-to-base and transmural gradients of relaxation that are believed to play important roles in active diastolic restoration of the LV and creation of forces required for diastolic suction (Sengupta et al. 2006). Kinetic energy during systole is consumed not only for ejection but also for conversion to potential energy by means of twisting the subendocardial fibers. During isovolumic relaxation (IVR) phase, the twisted subendocardial fibers behave similar to a compressed coil that springs open while releasing the potential energy stored in the deformed matrix. This recoil is appreciated as a clockwise rotation of the LV apex. During this phase, the rapid base-to-apex blood flow reversal prepares LV to efficiently receive blood. This rapid reversal of flow and change in intracavitary pressure gradient may explain the physiologic relevance of early endocardial relaxation, expansion of the LV cavity from apex and subsequent opening of the mitral valve.

3.5.4 Early and Late Diastole

Postsystolic shortening of the subendocardium near the base continues until the onset of early diastolic filling. Although untwisting of the apex occurs predominantly during IVR phase, untwisting extend further into early diastole. Expansion of the LV cavity in diastole occurs in three sequential stages; first, expansion at the apex that coincides with the onset of IVR phase, then, the LV base expands during early diastolic filling, and completes through late diastolic filling. During both early diastole and late diastole, an asymmetric vortex forms across the mitral valve where its larger section is across the anterior leaflet and its smaller section is across the posterior leaflet (Kheradvar et al. 2011).

3.6 Effect of Left Ventricular Diseases on Vortex Formation

The forming transmitral vortex *in vivo* rapidly interacts with the LV wall and deforms into an asymmetric flow structure, as discussed earlier. As a result, changes either in LV shape or contractile capacity of the heart would alter intraventricular flow pattern and forming vortex. Figure 3.11 shows examples of LV vortex in different heart diseases. The gross differences between normal and abnormal intraventricular flow are often well evidenced by the *steady streaming* pattern as shown in Fig. 3.11. The steady streaming image represents the flow field averaged during a heartbeat and can be considered the hallmark for the ventricular circulation. From this type of image, the relative geometrical relation between inflow jet, outflow tract, and the circulatory pattern can be qualitatively inferred.

3.6.1 Vortex Formation in LV Systolic Dysfunction

In patients with dilated LV, the transmitral flow is directed toward the free wall (Fig. 3.12) and gives rise to a well-developed circular flow pattern that turns toward the septum and the outflow tract during diastole (Mohiaddin 1995). Mathematical

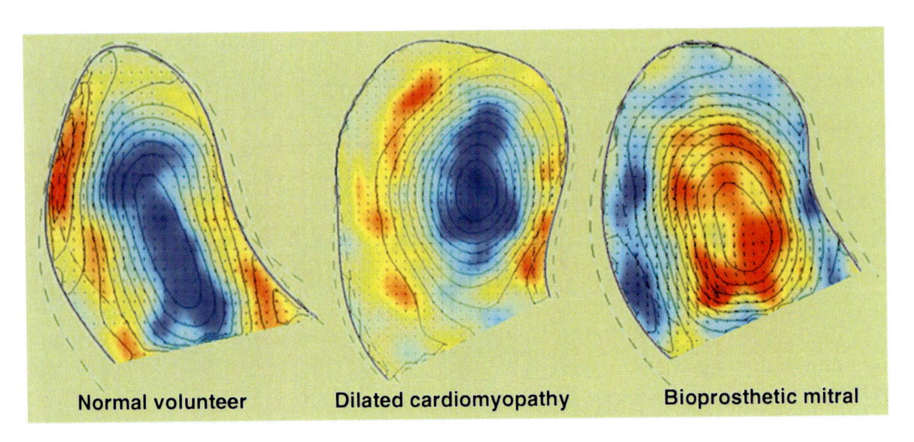

Fig. 3.11 Results from the steady streaming analysis in a normal volunteer (*left*), a patient with Dilated cardiomyopathy (*center*), and a patient with a bioprosthetic mitral valve (*right*). Pictures correspond to the A3C projection. *Color shading* represents circulatory pattern: *blue* is clockwise rotation and *red* is counterclockwise rotation. The flow pattern in normal left ventricles revealed a vortex moving in the basal-to-apical direction that initiated during diastole and dissipated during systole. In patients with DCM, the vortex shape was typically round and persistent. In most patients with prosthetic mitral valves, the vortex, in either the A4C or the A3C view, showed remarkable dissimilarity in the direction of rotation that was opposite to the normal subjects and more disturbed (Modified from Kheradvar et al. (2010))

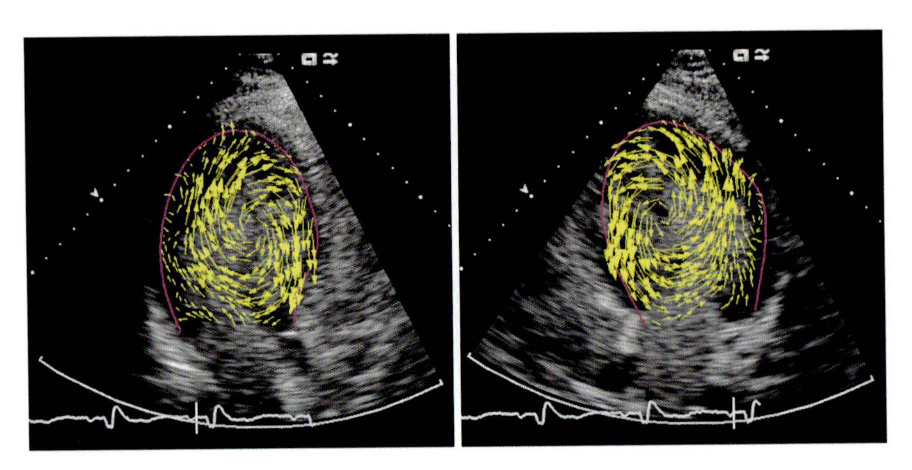

Fig. 3.12 Flow velocity at end-diastole in a patient with DCM, in A3C projection (*left*) and A4C projection (*right*). The typical DCM flow pattern revealed a round and persistent vortex that is fed during diastole and weakened during systole. In some subjects, however, the flow can appear more disturbed with a less coherent vortex structure

models have illustrated that the global LV systolic dysfunction would lead to delay in vortex detachment from mitral valve, reduced blood propagation velocity and longer stagnation near the apex (Baccani et al. 2002).

From pathophysiologic standpoint, the fluid stagnation may contribute to apical thrombus formation, which increases long-term mortality and morbidity. Optimal vortex formation is also critical for conserving kinetic energy; Bolger et al. demonstrated that about 16% of the kinetic energy of inflow was conserved at the end diastole in normal hearts compared to 5% in the dilated cardiomyopathy (DCM) hearts (Bolger et al. 2007). However, this observation was potentially due to the reduced direct flow in DCM, as no difference was found when results were normalized. Kheradvar et al. performed an *in vivo* study and compared the LV flow in normal volunteer, DCM patients, and patients with bioprosthetic mitral valve (Kheradvar et al. 2010). In DCM patients, the 2D representation of the LV vortex was more circular and symmetric compared to the elliptical and asymmetric shape observed in normal subjects (Kheradvar et al. 2010). In DCM hearts, the vortex appears stable and persistent. Based on these findings, it is yet unclear whether the development of a persistent smooth vortex in DCM hearts influences the progression of remodeling seen in these patients.

3.6.2 Vortex Formation in LV Regional Myocardial Dysfunction

Intraventricular flow is also influenced due to regional myocardial dysfunction. For example, infarcted segments of the LV can cause regional recirculation of blood flow not identified in healthy ventricles (Taylor et al. 1996). Beppu and colleagues demonstrated the existence of recirculatory flow areas in the basal LV during coronary ligation in dogs (Beppu et al. 1988). They found that the regional blood flow next to a dyskinetic wall exhibits little to no dynamic motion following the dyskinetic adjacent wall, resulting in a sort of flow stagnation region about the diskinetic region. In these patients, the transmitral jet is shorter and the vortex does not translate deep toward the LV apex. A recent mathematical model by Domenichini and Pedrizzetti also demonstrated that the presence of an anterior-inferior wall infarction may lead to shortening and weakening of the transmitral jet (Domenichini and Pedrizzetti 2011). A region of stagnant flow was found near the apex adjacent to the ischemic wall. Sometimes, the flow stagnation region adjacent to an infarcted segment can be detected based on existence of recirculatory flow area, which is mainly observed in diastasis when the rapid inflow is absent (Fig. 3.13). However, these recirculatory areas should not be confused with the transmitral vortex, as it is mostly a stagnant bubble adjacent to a dyskinetic wall that is not affected by the transmitral jet.

3.6.3 Hypertrophic Left Ventricle and Diastolic Dysfunction

Flow visualization has been used for understanding the mechanism of outflow tract obstruction in patients with hypertrophic cardiomyopathy. In the normal LV, formation of transmitral vortex helps maintaining the mitral leaflets closer to the posterior wall and assists directing the upcoming systolic stream toward the outflow tract. However, in hypertrophic cardiomyopathy, anterior displacement of the papillary muscles moves the entire mitral apparatus adjacent to the outflow tract, which

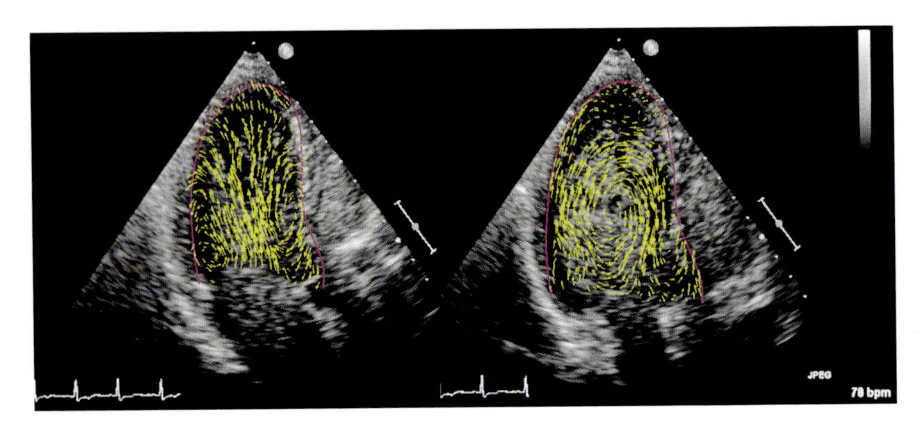

Fig. 3.13 Blood flow velocity during systole (*left*) and at diastasis (*right*) in a patient with apical infarction. The transmitral jet is weak and hardly reaches the apical region occupied by stagnant blood. The local recirculatory area next to the infarcted apical region is noticeable during diastasis with opposite rotational direction with respect to the transmitral vortex

reverses the direction of the transmitral vortex and promotes the systolic anterior motion of the mitral leaflets due to creation of drag forces (Lefebvre et al. 1995).

3.6.4 Vortex Formation in Mitral Stenosis

The fundamental pathophysiologic process underlying the mitral stenosis (MS) is partial mechanical obstruction of the LV inflow at the level of the mitral valve. Reduction of the mitral valve orifice area influences the transmitral flow to become increasingly dependent on a higher-than-normal pressure gradient between the LA and LV. The higher transmitral pressure gradient leads to several hemodynamic consequences such as elevated left atrial pressure, pulmonary artery hypertension with secondary effects on the pulmonary vasculature and the right heart. Mitral stenosis also results in LV dysfunction in affected patients while there is controversy on the etiology of dysfunction in patients with isolated MS. Evidence suggests that LV contractility in majority of patients with isolated MS appears normal or only slightly impaired (Carabello 2005). However, several studies reported impaired systolic function (Choi et al. 1995; Snyder et al. 1994; Surdacki et al. 1996; Mohan and Arora 1997) in addition to LV diastolic dysfunction (Dray et al. 2008; Paulus et al. 1992; Özer et al. 2004; Sengupta et al. 2004) in these patients.

Choi et al. evaluated 36 symptomatic patients with severe MS who were referred for hemodynamic evaluation prior to mitral valve surgery or balloon valvotomy (Choi et al. 1995). They reported subnormal EF (<45%) in more than 50% of these subjects. Paulus et al. used micromanometer recordings to evaluate early diastolic negative intraventricular pressures in MS patients and demonstrated loss of early diastolic suction effect in LV due to diastolic dysfunction (Paulus et al. 1992). Even

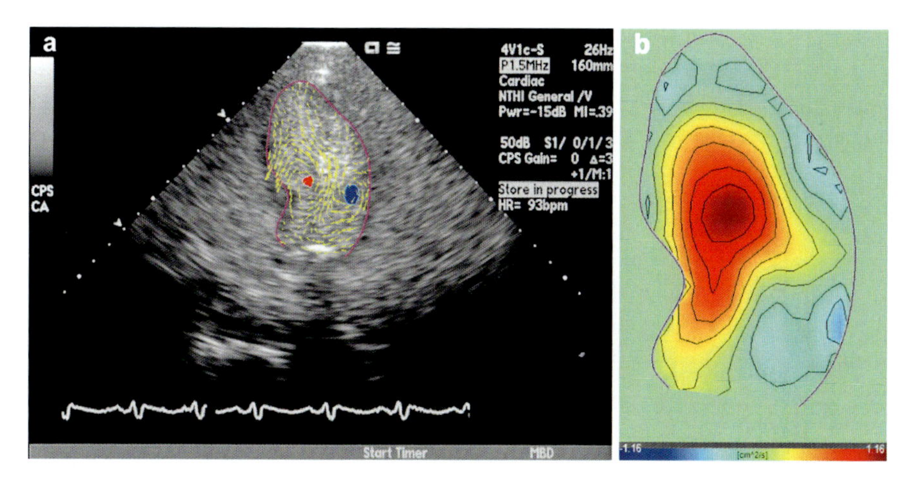

Fig. 3.14 Transmitral vortex formation in a patient with severe mitral stenosis. (**a**) Blood flow velocity at early-diastole superimposed by the transmitral vortex; blue and red represent opposite rotational directions. (**b**) The gross intraventrcicular flow pattern is evidenced by steady streaming pattern

in subjects whose EF was reported within the normal range based on standard echocardiography, LV dysfunction can be detected by more accurate methods such as tissue Doppler echocardiography or Strain Rate Imaging (SRI) (Özer et al. 2004; Özdemir et al. 2002; Dogan et al. 2006).

In the past, mechanical obstruction to the LV inflow was considered the main reason for LV dysfunction in MS patients (Hugenholtz et al. 1962). More recent studies proposed several other possible mechanisms including impaired diastolic filling, impaired contractility, excessive LV afterload, rigidity and fixation of the posterobasal LV myocardium from scarring or inflammation, altered interventricular interaction, or a combination of these factors (Mangoni et al. 2002; Klein and Carroll 2006). Klein and Carroll proposed a patient-specific and multifactorial etiology for impaired LV function in MS patients, and concluded that the diagnosis of LV dysfunction in these patients is difficult mainly due to subjectivity and load-dependence of echocardiography (Klein and Carroll 2006).

Vortex Formation Time index (VFT), as previously described, is a measure for efficiency of intraventricular blood transport from LA to LV. Kheradvar et al. in an *in vitro* study demonstrated that the diameter of the mitral valve annulus has a significant impact on transmitral flow efficiency and recoil of the mitral annulus (Kheradvar and Gharib 2007; Kheradvar et al. 2007). They showed that in a mitral valve with normal diameter, the maximal mitral annulus velocity as well as the maximal recoil force occurs about the onset of vortex ring pinch-off from the originating jet, indicating that the efficient blood transport mostly happens due to formation of the vortex ring. Recent work has suggested that severe MS may result in a small vortex with trailing jet that dissolves prior to detachment from transmitral jet (Fig. 3.14). Absence of a well-formed vortex, and dissipation of the stored energy may necessitate an increase in physical work generated by the cardiac muscle and lower efficiency of the heart pump (Jiamsripong et al. 2009). Clinical studies suggest that in patients with moderate to severe MS, the calculated value of VFT is much higher than its optimal range, which implies that the vortex never pinches off and is followed by a disordered wake pattern.

3.6.5 Vortex Formation Time Index and Percutaneous Balloon Mitral Valvotomy

In the absence of left atrial thrombus in a patient with favorable valve morphology, Percutaneous Balloon Mitral Valvotomy (PBMV) is the method of choice for majority of symptomatic patients with moderate-to-severe MS (Nishimura et al. 2008; Bonow et al. 2008). Meticulous assessment of the mitral valve anatomy by echocardiography is essential with regard to the feasibility and safety of PBMV. The most widely used echocardiographic tool for this purpose is the Wilkins score, which incorporates valve mobility, calcification, thickening, and the degree of subvalvular fusion (Wilkins et al. 1988). A mitral valve with a score less than eight or nine with moderate mitral regurgitation is deemed the best candidate for PBMV (Nobuyoshi et al. 2009).

To ensure the success of the PBMV procedure, a tangible measure of success needs to be defined. The most common indicator currently used is the mitral valve area (MVA) that is measured by direct tracing of the mitral orifice (planimetry), on a parasternal short-axis view (Baumgartner et al. 2009). However, this method may not be feasible even by experienced echocardiographers if there is a poor acoustic window or severe distortion of valve anatomy (Bonow et al. 2008). Doppler echocardiography is also used for indirect measurement of the valve area by pressure half-time method, and for assessment of the hemodynamic severity of the MS by measuring the mean transmitral pressure gradient (Baumgartner et al. 2009). Yet, pressure half-time method may be inaccurate in case of the LA or LV compliance abnormalities, in patients with associated aortic regurgitation, and in patients who have had mitral valvotomy (Bonow et al. 2008). Transmitral pressure gradient, which also has prognostic value, may not be the best option to assess the severity of MS due to its dependency on the mitral valve area, as well as the heart rate, cardiac output, and associated MR (Baumgartner et al. 2009). Three-dimensional echocardiography (Zamorano and de Agustín 2009; de Agustin et al. 2007; Messika-Zeitoun et al. 2007) along with indices (Rifaie et al. 2009) such as Mitral Leaflet Separation (MLS) index (Fisher et al. 1979; Seow et al. 2006; Holmin et al. 2007) or degree of commissural opening after PBMV (Messika-Zeitoun et al. 2009) have been suggested as possible alternative methods for assessment of PBMV success.

Very recently, the concept of VFT has been used for outcome analysis of PBMV. The LV hemodynamic of 8 patients prior to and after PBMV procedure was studied using echocardiography (Ghafourian et al. 2011). VFT significantly declined from a median baseline of 19.6 (7.9–25.5) to 7.9 (1.0–23.4) after PBMV ($P=0.02$). A statistically significant correlation was found between VFT and DT pre- ($r=0.89$, $p=0.007$) and post-PBMV ($r=0.83$, $p=0.01$). Additionally, the good correlation between VFT and DT pre- and post- PBMV, along with the independency of VFT from mitral valve area may suggest VFT as a useful tool for outcome analysis of PBMV. This is an early result and further studies are required to draw a firm conclusion.

References

Anderson RH. Clinical anatomy of the aortic root. Heart. 2000;84(6):670–3.

Andrew P. Diastolic heart failure demystified. Chest. 2003;124(2):744–53.

Baccani B et al. Fluid dynamics of the left ventricular filling in dilated cardiomyopathy. J Biomech. 2002;35(5):665–71.

Baumgartner H et al. Echocardiographic assessment of valve stenosis: EAE/ASE recommendations for clinical practice. Eur J Echocardiogr. 2009;10(1):1–25.

Bella JN et al. Mitral ratio of peak early to late diastolic filling velocity as a predictor of mortality in middle-aged and elderly adults: the strong heart study. Circulation. 2002;105(16):1928–33.

Bellhouse BJ. Fluid mechanics of a model mitral valve and left ventricle. Cardiovasc Res. 1972;6(2):199–210.

Bellhouse BJ, Talbot L. The fluid mechanics of the aortic valve. J Fluid Mech Digit Arch. 1969;35(04):721–35.

Beppu S et al. Abnormal blood pathways in left ventricular cavity in acute myocardial infarction. Experimental observations with special reference to regional wall motion abnormality and hemostasis. Circulation. 1988;78(1):157–64.

Berdajs D et al. Annulus fibrosus of the mitral valve: reality or myth. J Card Surg. 2007;22(5):406–9.

Bleasdale RA, Frenneaux MP. Prognostic importance of right ventricular dysfunction. Heart. 2002;88:323–4.

Boldt T, Andersson S, Eronen M. Outcome of structural heart disease diagnosed in utero. Scand Cardiovasc J. 2002;36(2):73–9.

Bolger AF, Heiberg E, Karlsson M, Wigström L, Engvall J, Sigfridsson A, et al. Transit of blood flow through the human left ventricle mapped by cardiovascular magnetic resonance. J Cardiovasc Magn Reson. 2007;9(5):741–7.

Bolzon G, Zovatto L, Pedrizzetti G. Birth of three-dimensionality in a pulsed jet through a circular orifice. J Fluid Mech. 2003;493(−1):209–18.

Bonow RO et al. 2008 Focused update incorporated into the ACC/AHA 2006 guidelines for the management of patients with valvular heart disease: a report of the American College of Cardiology/American Heart Association Task Force on practice guidelines (Writing Committee to revise the 1998 guidelines for the management of patients with valvular heart disease) endorsed by the Society of Cardiovascular Anesthesiologists, Society for Cardiovascular Angiography and Interventions, and Society of Thoracic Surgeons. J Am Coll Cardiol. 2008;52(13):e1–142.

Bonser RS et al. Surgical anatomy of the mitral and tricuspid valve, in mitral valve surgery. London: Springer; 2011. p. 3–19.

Buckberg GD et al. Active myocyte shortening during the 'isovolumetric relaxation' phase of diastole is responsible for ventricular suction; 'systolic ventricular filling'. Eur J Cardiothorac Surg. 2006;29(Suppl_1):S98–106.

Burgess MI et al. Comparison of echocardiographic markers of right ventricular function in determining prognosis in chronic pulmonary disease. J Am Soc Echocardiogr. 2002;15(6):633–9.

Campbell KB, Chandra M. Functions of stretch activation in heart muscle. J Gen Physiol. 2006;127(2):89–94.
Carabello BA. Modern management of mitral stenosis. Circulation. 2005;112(3):432–7.
Carlhall C et al. Contribution of mitral annular excursion and shape dynamics to total left ventricular volume change. Am J Physiol Heart Circ Physiol. 2004;287(4):H1836–41.
Chen L, Yin FCP, May-Newman K. The structure and mechanical properties of the mitral valve leaflet-strut chordae transition zone. J Biomech Eng. 2004;126(2):244–51.
Choi BW et al. Left ventricular systolic dysfunction diastolic filling characteristics and exercise cardiac reserve in mitral stenosis. Am J Cardiol. 1995;75(7):526–9.
da Vinci L. Quademi d'Anatomica II, 1513;9.
Dabiri JO, Gharib M. Fluid entrainment by isolated vortex rings. J Fluid Mech. 2004;511: 311–31.
D'Alonzo GE et al. Survival in patients with primary pulmonary hypertension: results from a National Prospective Registry. Ann Intern Med. 1991;115(5):343–9.
Davis JS et al. The overall pattern of cardiac contraction depends on a spatial gradient of myosin regulatory light chain phosphorylation. Cell. 2001;107(5):631–41.
de Agustin JA et al. The use of three-dimensional echocardiography for the evaluation of and treatment of mitral stenosis. Cardiol Clin. 2007;25(2):311–8.
de Groote P et al. Right ventricular ejection fraction is an independent predictor of survival in patients with moderate heart failure. J Am Coll Cardiol. 1998;32(4):948–54.
DeGroff CG et al. Flow in the early embryonic human heart. Pediatr Cardiol. 2003;24(4): 375–80.
Dincer I, Kumbasar D, Nergisoglu G, Atmaca Y, Kutlay S, Akyurek O, et al. Assessment of left ventricular diastolic function with Doppler tissue imaging: effects of preload and place of measurements. Int J Cardiovasc Imaging. 2002;18(3):155–60.
Dogan S et al. Prediction of subclinical left ventricular dysfunction with strain rate imaging in patients with mild to moderate rheumatic mitral stenosis. J Am Soc Echocardiogr. 2006;19(3): 243–8.
Dokainish H et al. Usefulness of new diastolic strain and strain rate indexes for the estimation of left ventricular filling pressure. Am J Cardiol. 2008;101(10):1504–9.
Domenichini F, Pedrizzetti G. Intraventricular vortex flow changes in the infarcted left ventricle: numerical results in an idealised 3D shape. Comput Methods Biomech Biomed Engin. 2011;14(1):95–101.
Domenichini F, Pedrizzetti G, Baccani B. Three-dimensional filling flow into a model left ventricle. J Fluid Mech. 2005;539:179–98.
Dray N, Balaguru D, Pauliks L. Abnormal left ventricular longitudinal wall motion in rheumatic mitral stenosis before and after balloon valvuloplasty: a strain rate imaging study. Pediatr Cardiol. 2008;29(3):663–6.
Fadlun EA et al. Combined immersed-boundary finite-difference methods for three-dimensional complex flow simulations. J Comput Phys. 2000;161(1):35–60.
Firstenberg MS, Greenberg NL, Main ML, Drinko JK, Odabashian JA, Thomas JD, et al. Determinants of diastolic myocardial tissue Doppler velocities: influences of relaxation and preload. J Appl Physiol. 2001;90(1):299–307.
Fisher ML, Parisi AF, Plotnick GD, DeFelice CE, Carliner NH, Fortuin NJ. Assessment of severity of mitral stenosis by echocardiographic leaflet separation. Arch Intern Med. 1979;139(4): 402–6.
Forouhar AS et al. The embryonic vertebrate heart tube is a dynamic suction pump. Science. 2006;312(5774):751–3.
Galderisi M. Diastolic dysfunction and diastolic heart failure: diagnostic, prognostic and therapeutic aspects. Cardiovasc Ultrasound. 2005;3(9):1–14.
Gharib M, Rambod E, Shariff K. A universal time scale for vortex ring formation. J Fluid Mech. 1998;360:121–40.
Gharib M, Rambod E, Kheradvar A, Sahn DJ, Dabiri JO. Optimal vortex formation as an index of cardiac health. Proc Natl Acad Sci USA. 2006;103(16):6305–8.

Ghafourian K, Falahatpisheh A, Goldstein SA, Pichard AD, Kheradvar A. Outcome analysis of percutaneous balloon mitral valvotomy through vortex formation time index. Circulation 2011;124:A13854

Graham Jr TP et al. Long-term outcome in congenitally corrected transposition of the great arteries: a multi-institutional study. J Am Coll Cardiol. 2000;36(1):255–61.

Gruber PJ, Epstein JA. Development gone awry: congenital heart disease. Circ Res. 2004;94(3):273–83.

Hamburger V, Hamilton H. A series of normal stages in the development of the chick embryo. J Morphol. 1951;88:49–92.

Holmin C et al. Mitral leaflet separation index: a new method for the evaluation of the severity of mitral stenosis? Usefulness before and after percutaneous mitral commissurotomy. J Am Soc Echocardiogr. 2007;20(10):1119–24.

Hong G-R et al. Characterization and quantification of vortex flow in the human left ventricle by contrast echocardiography using vector particle image velocimetry. J Am Coll Cardiol Imaging. 2008;1(6):705–17.

Hove JR et al. Intracardiac fluid forces are an essential epigenetic factor for embryonic cardiogenesis. Nature. 2003;421(6919):172–7.

Hugenholtz PG, Ryan TJ, Stein SW, Abelmann WH. The spectrum of pure mitral stenosis. Hemodynamic studies in relation to clinical disability. Am J Cardiol. 1962;10:773–84.

Jenkins C, Bricknell K, Marwick TH. Use of real-time three-dimensional echocardiography to measure left atrial volume: comparison with other echocardiographic techniques. J Am Soc Echocardiogr. 2005;18(9):991–7.

Jiamsripong P et al. Impact of acute moderate elevation in left ventricular afterload on diastolic transmitral flow efficiency: analysis by vortex formation time. J Am Soc Echocardiogr. 2009;22(4):427–31.

Jimenez JH et al. Effects of a saddle shaped annulus on mitral valve function and chordal force distribution: an in vitro study. Ann Biomed Eng. 2003;31(10):1171–81.

Jones EAV et al. Measuring hemodynamic changes during mammalian development. Am J Physiol Heart Circ Physiol. 2004;287(4):H1561–9.

Karlsson MO et al. Mitral valve opening in the ovine heart. Am J Physiol Heart Circ Physiol. 1998;274(2):H552–63.

Keren G, Sonnenblick EH, LeJemtel TH. Mitral anulus motion. Relation to pulmonary venous and transmitral flows in normal subjects and in patients with dilated cardiomyopathy. Circulation. 1988;78(3):621–9.

Kevin LG, Barnard M. Right ventricular failure. Contin Educ Anaesth Crit Care Pain. 2007;7(3):89–94.

Kheradvar A. Correlation between transmitral vortex formation and mitral valve's leaflet length. Circulation. 2010;122(21):A20561.

Kheradvar A, Gharib M. Influence of ventricular pressure drop on mitral annulus dynamics through the process of vortex ring formation. Ann Biomed Eng. 2007;35(12):2050–64.

Kheradvar A, Gharib M. On mitral valve dynamics and its connection to early diastolic flow. Ann Biomed Eng. 2009;37(1):1–13.

Kheradvar A, Kasalko J, Johnson D, Gharib M An in vitro study of changing profile heights in mitral bioprostheses and their influence on flow. ASAIO J. 2006;52(1):34–8.

Kheradvar A, Milano M, Gharib M. Correlation between vortex ring formation and mitral annulus dynamics during ventricular rapid filling. ASAIO J. 2007;53(1):8–16.

Kheradvar A, Assadi R, Jutzy KR, Bansal R Transmitral vortex formation: a reliable indicator for pseudonormal diastolic dysfunction. J Am Coll Cardiol. 2008;51:A104.

Kheradvar A, Houle H, Pedrizzetti G, Tonti G, Belcik T, Ashraf M, Lindner JR, Gharib M, Sahn DJ Echocardiographic particle image velocimetry: a novel technique for quantification of left ventricular blood vorticity pattern. J Am Soc Echocardiogr. 2010;23(1):86–94.

Kheradvar A, Assadi R, Falahatpisheh A, Sengupta PP. Assessment of transmitral vortex formation in patients with diastolic dysfunction, J Am Soc Echocardiogr. 2011 in press.

Kheradvar A, Falahatpisheh A. The effects of dynamic saddle annulus and leaflet length on transmitral flow pattern and leaflet stress of a bi-leaflet bioprosthetic mitral valve. J Heart Valve Dis. 2012 in press.

Khouri SJ, Maly GT, Suh DD, Walsh TE. A practical approach to the echocardiographic evaluation of diastolic function. J Am Soc Echocardiogr. 2004;17(3):290–7.

Kilner PJ et al. Asymmetric redirection of flow through the heart. Nature. 2000;404(6779): 759–61.

Kim WY et al. Two-dimensional mitral flow velocity profiles in pig models using epicardial echo Doppler cardiography. J Am Coll Cardiol. 1994;24(2):532–45.

Kim WY et al. Left ventricular blood flow patterns in normal subjects: a quantitative analysis by three-dimensional magnetic resonance velocity mapping. J Am Soc Echocardiogr. 1995;26(1):224–38.

Klein AJ, Carroll JD. Left ventricular dysfunction and mitral stenosis. Heart Fail Clin. 2006;2(4):443–52.

Krueger PS, Gharib M. The significance of vortex ring formation to the impulse and thrust of a starting jet. Phys Fluids. 2003;15(5):1271–81.

La Vecchia L et al. Reduced right ventricular ejection fraction as a marker for idiopathic dilated cardiomyopathy compared with ischemic left ventricular dysfunction. Am Heart J. 2001;142(1):181–9.

Lefebvre XP, He S, Levine RA, Yoganathan AP. Systolic anterior motion of the mitral valve in hypertrophic cardiomyopathy: an in vitro pulsatile flow study. J Heart Valve Dis. 1995;4(4): 422–38.

Levine RA et al. Echocardiographic measurement of right ventricular volume. Circulation. 1984;69(3):497–505.

Levine RA et al. Three-dimensional echocardiographic reconstruction of the mitral valve, with implications for the diagnosis of mitral valve prolapse. Circulation. 1989;80(3):589–98.

Liu A et al. Finite element modeling of blood flow-induced mechanical forces in the outflow tract of chick embryonic hearts. Comput Struct. 2007;85(11–14):727–38.

Mangoni AA et al. Outcome following mitral valve replacement in patients with mitral stenosis and moderately reduced left ventricular ejection fraction. Eur J Cardiothorac Surg. 2002;22(1):90–4.

Mangual J, Domenichini F, Pedrizzetti G. 3D echocardiographic assessment of right ventricular flow pattern. In: Euromech 529 – cardiovascular fluid mechanics. Cagliari; 2011.

McAlpine WA. Heart and coronary arteries: an anatomical atlas for clinical diagnosis, radiological investigation, and surgical treatment. New York: Springer Verlag; 1975.

Mehta SR et al. Impact of right ventricular involvement on mortality and morbidity in patients with inferior myocardial infarction. J Am Coll Cardiol. 2001;37(1):37–43.

Messika-Zeitoun D et al. Three-dimensional evaluation of the mitral valve area and commissural opening before and after percutaneous mitral commissurotomy in patients with mitral stenosis. Eur Heart J. 2007;28(1):72–9.

Messika-Zeitoun D et al. Impact of degree of commissural opening after percutaneous mitral commissurotomy on long-term outcome. J Am Coll Cardiol Imaging. 2009;2(1):1–7.

Mohan JC, Arora R. Effects of atrial fibrillation on left ventricular function and geometry in mitral stenosis. Am J Cardiol. 1997;80(12):1618–20.

Mohiaddin RH. Flow patterns in the dilated ischemic left ventricle studied by MR imaging with velocity vector mapping. J Magn Reson Imaging. 1995;5(5):493–8.

Moorman AFM, Soufan AT, Hagoort J, De Boer PAJ, Christoffels VM. Development of the building plan of the heart. Ann NY Acad Sci. 2004;1015:171–81.

Najos-Valencia O et al. Determinants of tissue Doppler measures of regional diastolic function during dobutamine stress echocardiography. Am Heart J. 2002;144(3):516–23.

Nakamura S et al. Right ventricular ejection fraction during exercise in patients with recent myocardial infarction: Effect of the interventricular septum. Am Heart J. 1994;127(1):49–55.

Nishimura RA, Tajik AJ. Evaluation of diastolic filling of left ventricle in health and disease: Doppler echocardiography is the clinicianâ€™s Rosetta stone. J Am Coll Cardiol. 1997;30(1): 8–18.

Nishimura RA et al. ACC/AHA 2008 guideline update on valvular heart disease: focused update on infective endocarditis: a report of the American College of Cardiology/American Heart Association Task Force on practice guidelines endorsed by the Society of Cardiovascular Anesthesiologists, Society for Cardiovascular Angiography and Interventions, and Society of Thoracic Surgeons. J Am Coll Cardiol. 2008;52(8):676–85.

Nobuyoshi M et al. Percutaneous balloon mitral valvuloplasty: a review. Circulation. 2009;119(8):e211–9.

Ogunyankin KO et al. Validity of revised Doppler echocardiographic algorithms and composite clinical and angiographic data in diagnosis of diastolic dysfunction. Echocardiography. 2006;23(10):817–28.

Oh JK et al. The noninvasive assessment of left ventricular diastolic function with two-dimensional and Doppler echocardiography. J Am Soc Echocardiogr. 1997;10(3):246–70.

Ommen SR et al. Clinical utility of Doppler echocardiography and tissue Doppler imaging in the estimation of left ventricular filling pressures: a comparative simultaneous Doppler-catheterization study. Circulation. 2000;102(15):1788–94.

Özdemir K et al. Analysis of the myocardial velocities in patients with mitral stenosis. J Am Soc Echocardiogr. 2002;15(12):1472–8.

Özer N et al. Left ventricular long-axis function is reduced in patients with rheumatic mitral stenosis. Echocardiography. 2004;21(2):107–12.

Pasipoularides A et al. Diastolic right ventricular filling vortex in normal and volume overload states. Am J Physiol Heart Circ Physiol. 2003;284(4):H1064–72.

Paulus WJ, Vantrimpont PJ, Rousseau MF. Diastolic function of the nonfilling human left ventricle. J Am Coll Cardiol. 1992;20(7):1524–32.

Peacock JA. An in vitro study of the onset of turbulence in the sinus of Valsalva. Circ Res. 1990;67(2):448–60.

Pedrizzetti G, Domenichini F. Nature optimizes the swirling flow in the human left ventricle. Phys Rev Lett. 2005;95(10):108101.

Pedrizzetti G, Domenichini F, Tonti G. On the left ventricular vortex reversal after mitral valve replacement. Ann Biomed Eng. 2010;38(3):769–73.

Peskin CS. The fluid dynamics of heart valves: experimental, theoretical, and computational methods. Annu Rev Fluid Mech. 1982;14(1):235–59.

Peskin CS, Wolfe AW. The aortic sinus vortex. Fed Proc. 1978;37(14):2784–92.

Peskin CS, McQueen DM, et al. Fluid dynamics of the heart and its valves. In: Othmer HG, editor. Case studies in mathematical modeling: ecology, physiology, and cell biology. Englewood Cliffs: Prentice-Hall; 1996. p. 309–37.

Petrie MC et al. Poor concordance of commonly used echocardiographic measures of left ventricular diastolic function in patients with suspected heart failure but preserved systolic function: is there a reliable echocardiographic measure of diastolic dysfunction? Heart. 2004;90(5):511–7.

Phoon CKL, Aristizábal O, Turnbull DH. Spatial velocity profile in mouse embryonic aorta and Doppler-derived volumetric flow: a preliminary model. Am J Physiol Heart Circ Physiol. 2002;283(3):H908–H916.

Prinzen FW et al. The time sequence of electrical and mechanical activation during spontaneous beating and ectopic stimulation. Eur Heart J. 1992;13(4):535–43.

Reul H, Talukder N, Muller EW. Fluid mechanics of the natural mitral valve. J Biomech. 1981;14(5):361–72.

Rifaie O et al. Can a novel echocardiographic score better predict outcome after percutaneous balloon mitral valvuloplasty? Echocardiography. 2009;26(2):119–27.

Robicsek F. Leonardo da Vinci and the sinuses of Valsalva. Ann Thorac Surg. 1991;52(2): 328–35.

Ryan LP et al. Description of regional mitral annular nonplanarity in healthy human subjects: a novel methodology. J Thorac Cardiovasc Surg. 2007;134(3):644–8.

Salgo IS et al. Effect of annular shape on leaflet curvature in reducing mitral leaflet stress. Circulation. 2002;106(6):711–7.

Santhanakrishnan A, Miller LA. Fluid dynamics of heart development. Cell Biochem Biophys. 2011;61:1.

Santhanakrishnan, A., Nguyen, N., Cox, J. G., and Miller, L. A. (2009) Flow within Models of the Vertebrate Embryonic Heart. Journal of Theoretical Biology, 259, 449–464.

Schillaci G et al. Prognostic significance of left ventricular diastolic dysfunction in essential hypertension. J Am Coll Cardiol. 2002;39(12):2005–11.

Selamet Tierney ES et al. Changes in left heart hemodynamics after technically successful in-utero aortic valvuloplasty. Ultrasound Obstet Gynecol. 2007;30(5):715–20.

Sengupta PP et al. Effects of percutaneous mitral commissurotomy on longitudinal left ventricular dynamics in mitral stenosis: quantitative assessment by tissue velocity imaging. J Am Soc Echocardiogr. 2004;17(8):824–8.

Sengupta PP et al. Biphasic tissue Doppler waveforms during isovolumic phases are associated with asynchronous deformation of subendocardial and subepicardial layers. J Appl Physiol. 2005;99(3):1104–11.

Sengupta PP et al. Apex-to-base dispersion in regional timing of left ventricular shortening and lengthening. J Am Coll Cardiol. 2006;47(1):163–72.

Sengupta PP, Khandheria BK, Korinek J, Jahangir A, Yoshifuku S, Milosevc I, et al. Left ventricular isovolumic flow sequence during sinus and paced rhythms: new insights from use of high-resolution Doppler and ultrasonic digital particle imaging velocimetry. J Am Coll Cardiol. 2007;49:899–908.

Seow S-C, Koh L-P, Yeo T-C. Hemodynamic significance of mitral stenosis: use of a simple, novel index by 2-dimensional echocardiography. J Am Soc Echocardiogr. 2006;19(1):102–6.

Shariff K, Leonard A. Vortex rings. Annu Rev Fluid Mech. 1992;24:U235–79.

Silverman NH, Hudson S. Evaluation of right ventricular volume and ejection fraction in children by two dimensional echocardiography. Pediatr Cardiol. 1983;4:197–204.

Snyder II RW, Lange RA, Willard JE, Glamann DB, Landau C, Negus BH, et al. Frequency, cause and effect on operative outcome of depressed left ventricular ejection fraction in mitral stenosis. Am J Cardiol. 1994;73(1):65–9.

Sohn DW et al. Assessment of mitral annulus velocity by Doppler tissue imaging in the evaluation of left ventricular diastolic function. J Am Coll Cardiol. 1997;30(2):474–80.

Surdacki A, Legutko J, Turek P, Dudek D, Zmudka K, Dubiel JS. Determinants of depressed left ventricular ejection fraction in pure mitral stenosis with preserved sinus rhythm. J Heart Valve Dis. 1996;5(1):1–9.

Taber LA, Yang M, Podszus WW. Mechanics of ventricular torsion. J Biomech. 1996;29(6): 745–52.

Taylor TW, Suga H, Goto Y, Okino H, Yamaguchi T. The effects of cardiac infarction on realistic three-dimensional left ventricular blood ejection. J Biomech Eng. 1996;118(1):106–10.

Tayyareci Y et al. Early detection of right ventricular systolic dysfunction by using myocardial acceleration during isovolumic contraction in patients with mitral stenosis. Eur J Echocardiogr. 2008;9(4):516–21.

Van Steenhoven AA, Van Dongen MEH. Model studies of the closing behaviour of the aortic valve. J Fluid Mech Digit Arch. 1979;90(01):21–32.

Vennemann P et al. In vivo micro particle image velocimetry measurements of blood-plasma in the embryonic avian heart. J Biomech. 2006;39(7):1191–200.

Wang Y, Dur O, Patrick MJ, Tinney JP, Tobita K, Keller BB, Pekkan K. Aortic arch morphogenesis and flow modeling in the chick embryo. Ann Biomed Eng. 2009;37(6):1069–81.

Whalley GA et al. Comparison of different methods for detection of diastolic filling abnormalities. J Am Soc Echocardiogr. 2005;18(7):710–7.

Wilkins GT et al. Percutaneous balloon dilatation of the mitral valve: an analysis of echocardiographic variables related to outcome and the mechanism of dilatation. Br Heart J. 1988;60(4):299–308.

Wilkins-Haug LE et al. In-utero intervention for hypoplastic left heart syndrome – a perinatologist's perspective. Ultrasound Obstet Gynecol. 2005;26(5):481–6.

Xie G-Y, Smith MD. Pseudonormal or intermediate pattern? J Am Coll Cardiol. 2002;39(11): 1796–8.

Zamorano J, de Agustín JA. Three-dimensional echocardiography for assessment of mitral valve stenosis. Curr Opin Cardiol. 2009;24(5):415–9. doi:10.1097/HCO.0b013e32832e165b.

Zehender M et al. Eligibility for and benefit of thrombolytic therapy in inferior myocardial infarction: focus on the prognostic importance of right ventricular infarction. J Am Coll Cardiol. 1994;24(2):362–9.

Zile MR, Brutsaert DL. New concepts in diastolic dysfunction and diastolic heart failure: part I: diagnosis, prognosis, and measurements of diastolic function. Circulation. 2002;105(11): 1387–93.

Chapter 4
Effect of Cardiac Devices and Surgery on Vortex Formation

Abstract In this chapter, the fluid dynamics of the artificial heart valves and ventricular assist devices are described. The effect of each device on cardiovascular vortex formation is comprehensively reviewed. At last, the unnatural vortices and their formation due to surgical procedures for correction of heart defects and anastomoses are discussed.

4.1 Vortex Formation in Presence of Bioprosthetic Heart Valves

Valvular heart disease is the next cardiac epidemic (d'Arcy et al. 2011). Currently, valvular heart disease is the third most common cause of heart problems in the United States. Although the accurate overall picture of the epidemiology of valvular heart disease has not yet established, the overall prevalence in the US is 2.5% with wide age-related variation from 0.7% in younger population to 13.3% in elderly population (Nkomo et al. 2006). More than 200,000 procedures are annually performed all over the world to replace natural heart valves that are severely damaged (Abu-Omar and Ratnatunga 2008). Replacement of dysfunctional valves markedly reduces the morbidity and mortality associated with the valve disease. Current options are mainly limited to mechanical or bioprosthetic valve types. Each of these valve groups has its advantages and limitations, and therefore, surgeons try to match the valve choice with the patient's particular clinical condition.

Mechanical heart valves (MHVs) tend to last longer than bioprosthetic heart valves (BHVs) due to their stronger composition, but they carry a greater long-term risk for thromboembolism that may lead to stroke and arterial thrombosis. Mechanical valves are generally used in younger patients because of durability. However, the lifelong need for anticoagulant medication is a major drawback to these valves (Senthilnathan et al. 1999). BHVs, in contrast, do not require anticoagulant medications due to their biocompatible surface and improved blood flow dynamics that minimizes red blood cell damage. Their lower risks of thrombogenicity

A. Kheradvar, G. Pedrizzetti, *Vortex Formation in the Cardiovascular System*,
DOI 10.1007/978-1-4471-2288-3_4,

Fig. 4.1 Streamlines of transmitral jet downstream a 25 mm Mitral Biocor™ by St. Jude Medical is obtained through high-speed particle image velocimetry. The leading symmetrical vortex ring in front of the jet can be observed. The symmetricity of the formed vortex is due to the circular opening of these valves (Modified from Kheradvar and Falahatpisheh 2012)

and superior hemodynamics, when compared to the mechanical valves, have given these valves remarkable advantages. Nevertheless, BHVs are made of animal tissues (either porcine valve leaflet or bovine pericardium), which undergo several chemical procedures to be suitable for implantation in the human heart. These xenograft heart valve prostheses are associated with immune reactions and progressive deterioration with limited durability (Apte et al. 2011; Stephens et al. 2010). The degeneration rate of BHVs is inversely related to the age of the patient at the time of implantation; accordingly, BHVs can be mainly recommended for patients aged 65 or older (Hoffmann et al. 2008).

Since BHVs are centrally opening valves, the pulsatile flow through them develops a symmetric vortex ring (Fig. 4.1), regardless of the position of the implant in the heart; the valves implanted at mitral and tricuspid positions result in an intraventricular vortex while valves implanted at aortic and pulmonary positions develop a vortex in valsavla sinuses and pulmonary trunk, respectively.

The physiologic vortex that develops through a natural heart valve helps mixing the blood, minimizes the residence time of blood in a chamber and effectively transfers the blood momentum from one chamber to the next. Ideal BHVs should maintain similar functionality as of a natural heart valve. Several parameters can affect the hemodynamics and the flow field downstream a valve. These parameters can be different for each valve depending on the valve's implanting location and functionality. Here, we discuss about the valvular hemodynamics at different positions.

4.1.1 Mitral Bioprosthetic Valves

Bioprosthetic valves are routinely used to replace stenotic or regurgitant natural mitral valves. Among those, Biocor™ from St. Jude Medical, Carpentier-Edwards Mitral Porcine and Carpentier-Edwards PERIMOUNT™ (CEP) series by Edwards Lifesciences are the most common valves currently being used for implantation. Investigators have previously shown that the height of the valve profile is a major contributor to the flow field downstream a centrally opening heart valve (Kheradvar et al. 2006; Leo et al. 2005). Kheradvar et al. compared the flow downstream four CEP mitral prototypes using particle image velocimetry (Fig. 4.2) (Kheradvar et al. 2006). It was found that increasing the atrial projection of a mitral bioprosthesis results in reduction of the magnitude of circulation (Γ) downstream the valve (Fig. 4.3). Based on the velocity field, they computed the particle residence time at the stagnation area near the valve. Particle residence time (T_p) denotes the time required for a designated particle to leave an area of interest. It provides quantitative information about flow stagnation in certain regions around a heart valve, and may correlate with flow-induced thrombogenicity of the device. Forming vortices may improve or aggravate the particle residence time depending on induction of recirculation zones.

Goetze et al. studied the short-term hemodynamic profile of the CEP pericardial bioprostheses using Doppler echocardiographic in 189 patients (Goetze et al. 2004). They found that the CEP mitral valves overall have favorable transvalvular pressure gradient across various sizes of valve, which is particularly critical for smaller valve sizes. However, currently no quantitative *in vivo* information exists on how vortex is formed through bioprosthetic mitral valves.

Kheradvar and Gharib (2009) found a strong logarithmic relationship ($R^2=0.94$) that describes the vortex formation time in terms of ratio of pressure drop

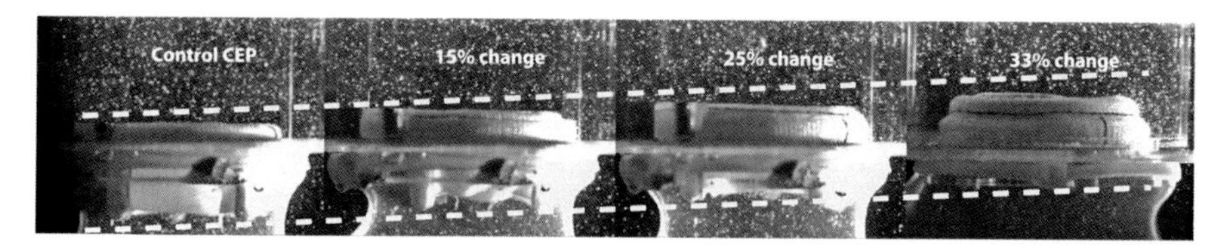

Fig. 4.2 Profile height changes in Perimount™ mitral valves. From left to right, control CEP valve, prototypes with 15%, 25% and 33% elevation in atrial side and the corresponding reduction in ventricular height (Modified from Kheradvar et al. (2006))

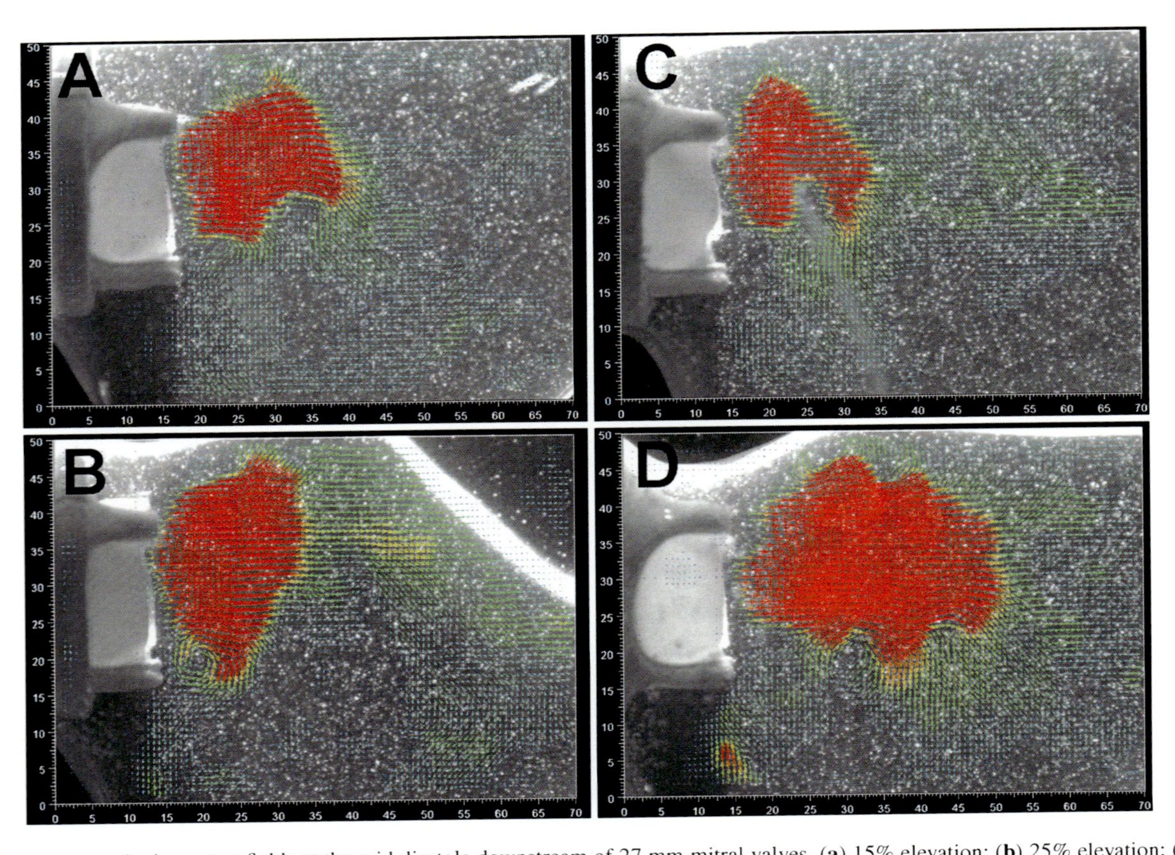

Fig. 4.3 Ventricular velocity vector fields at the mid diastole downstream of 27 mm mitral valves. (**a**) 15% elevation; (**b**) 25% elevation; (**c**) 33% elevation, and (**d**) control CEP valve. The cross-section of the vortex downstream the valves could be observed

time-constant (τ) to duration of the isovolumic relaxation time (*IVRT*) *in vitro* using a 27 mm CEP mitral valve:

$$VFT = 3.845Log\left(\frac{\tau}{IVRT}\right) + 9.394 \tag{4.1}$$

The relationship predicts that once the ratio of pressure drop time-constant to isovolumic relaxation time period increases, the vortex formation time is extended. This relationship shows the fate of the transmitral vortex and the onset of detachment from its jet is mostly determined based on the structural characteristics of the LV that define the diastolic pressure drop time-constant and the duration of isovolumic relaxation phase. However, size of the bioprosthetic valve can affect the transmitral jet profile during the formation process and influence the vortex formation time and pinch-off process artificially. This is similar to what naturally occurs during mitral stenosis in which the jet stroke ratio is extensively extended and may even prevent the roll-up and formation of the leading vortex ring (Sect 3.6.4).

4.1.2 Aortic Bioprosthetic Valves

The vortex that develops in the sinuses of valsalva helps mixing the blood, improves the coronary blood flow and facilitate the valve closure (Katayama et al. 2008; Van Steenhoven and Van Dongen 1979). These vortices have been studied by several researchers; Leonardo Da Vinci was the first who identified these flow structures (da Vinci 1513; Gharib et al. 2002). Peacock experimentally showed that the aortic sinus vortex can become weakly turbulent under simulated exercise conditions with turbulent intensities that are highest near the location of coronary ostia (Peacock 1990). He found that despite the transition to turbulence within the vortex, the mainstream aortic flow upstream the valve remains laminar (Peacock 1990). This feature helps maintaining efficient blood stream into aorta while providing adequate flow for the coronary arteries.

An ideal aortic prosthesis should generate a vortex in the sinuses of valsalva, feed the coronary arteries, while maintaining a laminar stream into the aortic arch similar to a natural aortic valve. It is expected that the centrally opening aortic bioprostheses provide such type of flow field due to their similarity to a native aortic valve. However, no experimental study has yet confirmed it.

4.2 Vortex Formation in Presence of Mechanical Heart Valves (MHVs)

Mechanical heart valves (MHV) are considered the better choice for patients under the age of 60 due to their remarkable durability (Rahimtoola 2010). The rigid leaflets of the MHVs pose an obstruction to the blood flow producing significant flow disturbances. These abnormal flow patterns provoke major clinical complications and dramatically affect the performance and longevity of the valve. Major problems associated with the valve hemodynamics are hemolysis (Shapira et al. 2009), thrombosis, thromboembolism, microbubbles and damage to the endothelial tissue.

Thromboembolism is a major factor that may contribute to stroke and other cardiovascular diseases (Skjelland et al. 2008). The most lethal form of thromboembolism is thrombotic prosthetic obstruction. Nowadays, the incidence of major embolism varies in the range of 0.6–4.3% per patient year, depending on the type of valve, antithrombotic therapy and the valve position (Zilla et al. 2008; Cannegieter et al. 1994). The patients require a mandatory life-long anticoagulant drug regimen, which exposes them to hemorrhage vulnerability. Thrombus formation and blood coagulation incidence are strongly related to shear induced platelet activation. According to hematologic studies (Raz et al. 2007; Morbiducci et al. 2009), platelet activation/aggregation is significantly greater under pulsatile or unsteady conditions, and is a function of the time-cumulative shear stress over individual platelet trajectories. In addition, flow separation, vortex shedding and recirculatory regions with long blood residence times, as often found in the wake of the MHV leaflets, provide optimal conditions for coagulation factors to mix and activate platelets to aggregate. Thus MHVs facilitate platelets activation through shear layers, platelet aggregation and thrombus formation through downstream wakes. These factors are even more enforced by the motion of the passive valve's leaflets, especially during valve closure, when extreme pressure drop and squeeze flow may result in cavitation and formation of emboli. Furthermore, elevated fluid stresses lead to endothelial cells damage in a manner that may lead to further adherence, activation, and aggregation of platelets (Yoganathan et al. 2005).

4.2.1 Study of MHVs' Hemodynamics

The importance of the flow field across MHVs has produced a large number of studies over the last four decades. The ultimate aim of these studies is to aid the design of optimal MHVs with reduced blood damage and coagulation potential through understanding valvular flow field. The problem is challenging from several standpoints:

a. The complex geometry of the atrium and the ventricle with large anatomical and physiological variations can be a serious valve design concern for mitral valve replacement. At the aortic position, the exact anatomic geometry of the aortic root and sinus after aortotomy, valve's orientation and the flexible curved aorta are crucial for the determination of local flow field.
b. The flow passing the valve is highly unsteady. The sources of unsteadiness are the natural vortex shedding from the leaflet tips, variations imposed by the time-dependent inflow, myocardial motion, and the motion of the valve's leaflets.
c. The turbulent behavior of the flow is unpredicted. The Reynolds number of blood flow across an MHV ranges from 600 to 6,000, indicating a rich mixture of small-scale nonlinear flow instabilities. The pulsatile flow increases the instability,

particularly during the decelerating phase. Such unpredictable transitional flow is challenging to either measure or model.

d. The passive motion of the valve's leaflets strongly depends on the instantaneous flow, and in turn, dramatically affects it, particularly during valve closure. At the mitral position, transmitral flow not only determines the motion of the valve's leaflets, but also affects the entire LV flow and even the aortic leaflets dynamics.
e. The local flow dynamics near the leaflets' hinges significantly influence the global flow dynamics, and in fact, depend on combination of hinge geometry and the ventricular flow conditions.
f. The blood is a non-Newtonian, two-phase, nonhomogeneous fluid. The nonlinear characteristics of blood contribute to the complexity of the problem, and require careful consideration when predicting the effect of local flow phenomena on platelets activation or valve's hinge.

The combination of the above factors makes the study of blood flow across MHVs an extremely challenging fluid dynamics' problem.

Studies performed *in vivo* addressing the blood flow across MHVs, exploit ultrasound Doppler velocimetry (Maire et al. 1994; Pop et al. 1989; Van Rijk-Zwikker et al. 1996), MRI (Machler et al. 2004) or Echocardiographic Particle Image Velocimetry (Faludi et al. 2010; Kheradvar et al. 2010) to reveal the influence of valve's geometry and orientation on ventricular flow. Studies performed *in vitro* use mainly Laser Doppler Velocimetry (Meyer et al. 2001; Kini et al. 2001; Simon et al. 2004; Travis et al. 2002; Schoephoerster and Chandran 1991) and Particle Image Velocimetry (PIV) (Krishnan et al. 2006; Kheradvar et al. 2006; Kini et al. 2001; Travis et al. 2001; Shu et al. 2004; Raz et al. 2002; Rambod et al. 2007; Querzoli et al. 2010; Mouret et al. 2005; Manning et al. 2008; Li et al. 2010; Knapp and Bertrand 2005; Kelly 2002; Kaminsky et al. 2007; Govindarajan et al. 2009a) downstream of MHV in controlled conditions. Other methods use the phase-contrast MRI (Kvitting et al. 2010) and dye washout (Goubergrits et al. 2008) to investigate specific parameters. Computational studies are mainly based on numerical simulations of the flow across the valve (Morbiducci et al. 2009; Sotiropoulos and Borazjani 2009; Avrahami et al. 2000; Nobili et al. 2008; Cheng et al. 2004; Ge et al. 2005) or on the coupled flow and leaflet interaction (de Tullio et al. 2011; Goubergrits et al. 2008; Bluestein et al. 2010; Einav et al. 2002). They aim to obtain the most detailed information about the flow variables, isolate parameters and design optimization. For further accuracy, the numerical calculations are often validated with experiments (Kelly 2002; Nobili et al. 2008; Cheng et al. 2004; Ge et al. 2005; Einav et al. 2002; Smadi et al. 2010).

4.2.2 *Types of MHVs*

Currently, there are two main types of MHVs in clinical use: Monoleaflet MHV (MMHV) and bileaflet MHV (BMHV). MMHVs were mainly implanted from 1975 through 1995. Since early 1990s, BMHVs have taken over the market. However, the studies based on implantation of both valve types revealed similar clinical performance; no significant differences were found in early mortality, long-term survival, or other valve-related complications. BMHVs exhibit a little higher rate of thrombosis and their delicate hinge mechanism tends to induce severe mechanical dysfunction (Kaminsky et al. 2007; Meuris et al. 2005; Wu et al. 2004). Both types of MHVs exhibit comparable mainstream flow characteristics. During forward flow, both valves develop strong shear layers near the edges of the leaflets, which shed large-scale Von Karman vortices (see Sect. 2.4) downstream the valve. The shed vortices are characterized by birth of turbulence and influence the flow farther away downstream the valve.

During the flow deceleration, a simultaneous development of vortices takes place in both valve types. These vortices are related to vorticity waves (Rosenfeld et al. 2002), which are essentially the inviscid mechanism that generates rotational flow due to residual kinetic energy following a sudden decrease in net flow. Presence of vorticity waves is typical in intermittent turbulent flow conditions. During the deceleration phases, the turbulent shear stresses are relatively low (Dasi et al. 2009), but the appearance of vortices in the wake of the valve provides optimal conditions that lead to further platelet aggregation and clotting reactions.

The valve closure onset is critical for both types of MHVs. The phase of leaflet closing lasts only about 10 ms (~ 1% of a cardiac cycle). Yet, it is believed to play a major role in the valve's pathophysiology. As the pressure gradient across the valve is inverted, the valve's leaflets move backwards to occlude the orifice with high leaflet tip velocities in the order of 5 m/s. During the valve closure, shear layers generated by backflow form strong Rankine vortices on the upstream side of the leaflets. It is well known that the pressure decreases toward the center of the Rankine vortices. The stronger the vortex, the lower the pressure at the center will be. Therefore, the low pressure that develops at the center of the tip vortices may even reach blood evaporation pressure. The evaporation only occurs during closure, and is strongly dependent on flow conditions and valve size (Hose et al. 2006). This situation is more complicated when MHVs are placed at the mitral position (Zimpfer et al. 2006). In addition, the slamming effect of the occluders against the ring leads to leaflets rebound and thus further pressure drop (Kini et al. 2001). When combined with closure squeeze flow, it can create favorable conditions for the inception of cavitation bubbles (Lim et al. 2003). Cavitation bubbles are attributed to pitting and erosion marks on the valves' leaflets, which have been suggested to potentially cause valve's failure (Eichler and Reul 2004; Lee et al. 2010). Although highly polished surfaces reduces the risk of leaflet erosion, collapse of cavitation bubbles is suspected to promote platelet activation and aggregation (Milo et al. 2003) and to damage the endothelial lining (Brujan 2009).

Once microbubbles' nuclei are formed, dissolved air gases easily release from blood in the low pressure zone. These gases may include Nitrogen (Rambod et al. 2007) or carbon dioxide (Brujan 2009). The released gases increase the bubbles' volume and maintain them when they leave the low-pressure zone. This bubble generation mechanism may explain the gas-filled bubbles emboli of the order of 200–300 μm in diameter with lifespan of several seconds that are observed by echocardiography (Johansen 2004). These microbubbles are frequently detected in the cerebral circulation of patients with MHVs and are attributed to damage the brain tissue, resulting in cognitive impairments and a decline in neurological function (Zimpfer et al. 2006; Deklunder et al. 1998; Telman et al. 2002). During backward flow, both valves produce strong upstream jets, significant pressure drop, large stresses and squeeze flow. Narrow gaps between the disc and the housing in both types of valves enable regurgitant flow even when the leaflets are closed. Since the jet velocities are very high, they induce critical turbulent stresses, which are even greater than those during forward flow.

The specific flow developed across each MHV is strongly dependent on valve type, position, design, orientation and physiologic waveform. There are several significant differences between the flow across MMHVs and BMHVs. BMHVs induce typical triple-jet stream (one main and two side jets), which is relatively symmetric. However, MMHVs induce asymmetric flow with a two-jet diagonal stream. Therefore, at the mitral position, MMHVs in the posterior orientation tend to preserve more natural vortical motion that dominates the LV flow (Faludi et al. 2010), facilitating ejection of blood during systole (Pedrizzetti et al. 2010). At the aortic position, MMHVs are considered to yield better coronary flow (Kleine et al. 2002; Bakhtiary et al. 2007). The MMHVs demonstrate a slightly lower regurgitant volume and less pressure drops than the BMHV design (Yoganathan et al. 2004). Conversely, BMHVs open a little faster than the MMHVs (Kaminsky et al. 2007).

4.2.3 Flow Across Mono-leaflet MHVs (MMHVs)

In MMHVs, two jets are emanated from the major and minor orifices of the tilting disk of the valve during forward flow (i.e. systole for the aortic valve and diastole for the mitral valve). The major orifice jet is larger and has a slightly higher velocity than the minor orifice jet. These jets induce two large tip vortices at the leaflets' edges. The shear layer originated at the minor orifice from the rear edge of the leaflet rolls-up in a large vortex that stays adjacent to the leaflet's wall, forming a growing stagnation region shielded by the leaflet (Fig. 4.4). The shear layer emanated from the large orifice rolls-up at the leading edge of the leaflet and sheds von Karman vortices downstream (Rosenfeld et al. 2002). The upper jet produces vortical flow in the third direction in the vicinity of the leaflet. Under physiological conditions, the peak velocities measured downstream the major and minor orifices are similar, and about 2 m/s, with high shear stresses (Yoganathan et al. 2004).

As the flow starts to decelerate, a backward pushing pressure develops and the tilting disc starts moving back passively. The sudden flow deceleration and disc motion cause the simultaneous development of large vortices downstream the valve due to a vorticity wave. The turbulent shear stresses during the deceleration phases are relatively low (Dasi et al. 2009), but the appearance of vortices in the wake of the valve provides optimal conditions for further platelet aggregation and for the clotting reactions to occur. When the pressure gradient across the valve is inverted, the occluder moves towards the closing position with leaflet tip velocities up to 5 m/s until it hits the valve's ring. During the valve closure, as the gap size between

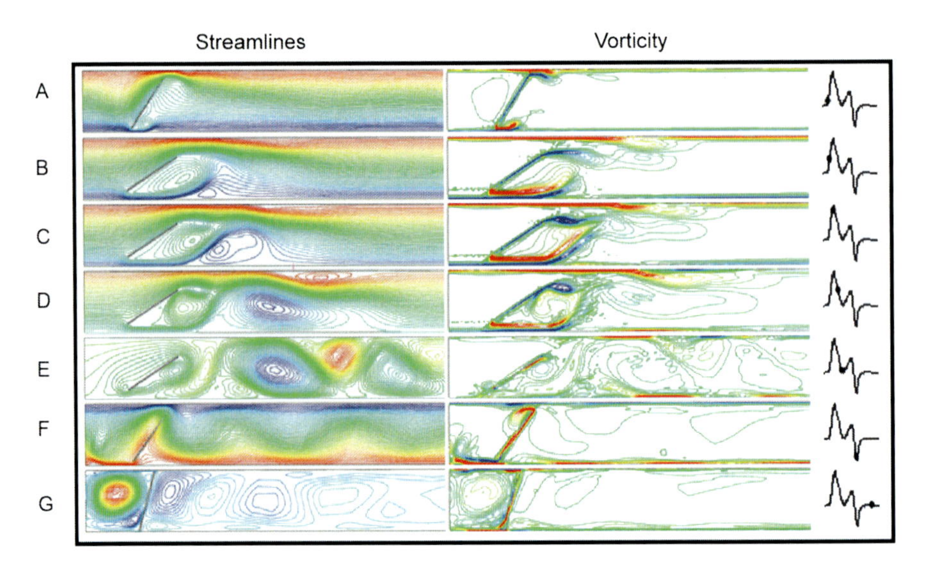

Fig. 4.4 Streamlines and vorticity of the time-dependent flow obtained through a numerical model of an MMHV in a straight channel (The images are adapted from Rosenfeld et al. (2002))

the leaflet and the valve housing decreases, the blood velocity in the narrow passages increases and reaches high values, inducing abnormal high shear stress (up to 18,000 dyn/cm^2, Meyer et al. 2001). These high velocities result in an increase in the backwards pressure drop across the valve once the strong shear layer generated from the large orifice during the valve closure rolls up and feeds a strong Rankine vortex on the upstream side of the leaflet (Avrahami et al. 2000; Li et al. 2008) (Fig. 4.5a). An additional pressure drop is measured at the center of the tip vortices, with pressure minima at the center of the larger orifice vortex (Fig. 4.5b), indicating favorable conditions for the inception of cavitation bubbles.

When the leaflet is closed (i.e. systole for the mitral valve and diastole for the aortic valve), the backward pressure-drop across the valve ensures the leaflet closure. Yet, due to the small gap between the leaflet and the support ring in the minor orifice, a small amount of regurgitant flow is often observed. Although the MMHVs' design is characterized with relatively lower regurgitant volumes than the BMHVs, the measured velocities are significantly (Meyer et al. 2001; Maymir et al. 1997) indicating that the thromboembolic and hemolytic potential of regurgitant flow is comparable to, if not greater than the forward flow.

4.2.4 *Orientation of MMHVs*

The velocity profiles and the magnitudes of turbulent shear stress downstream the MMHVs vary based on valve design, valve's opening angle and position/orientation of the valve (Maire et al. 1994; Chandran et al. 2006). For an MMHV at aortic

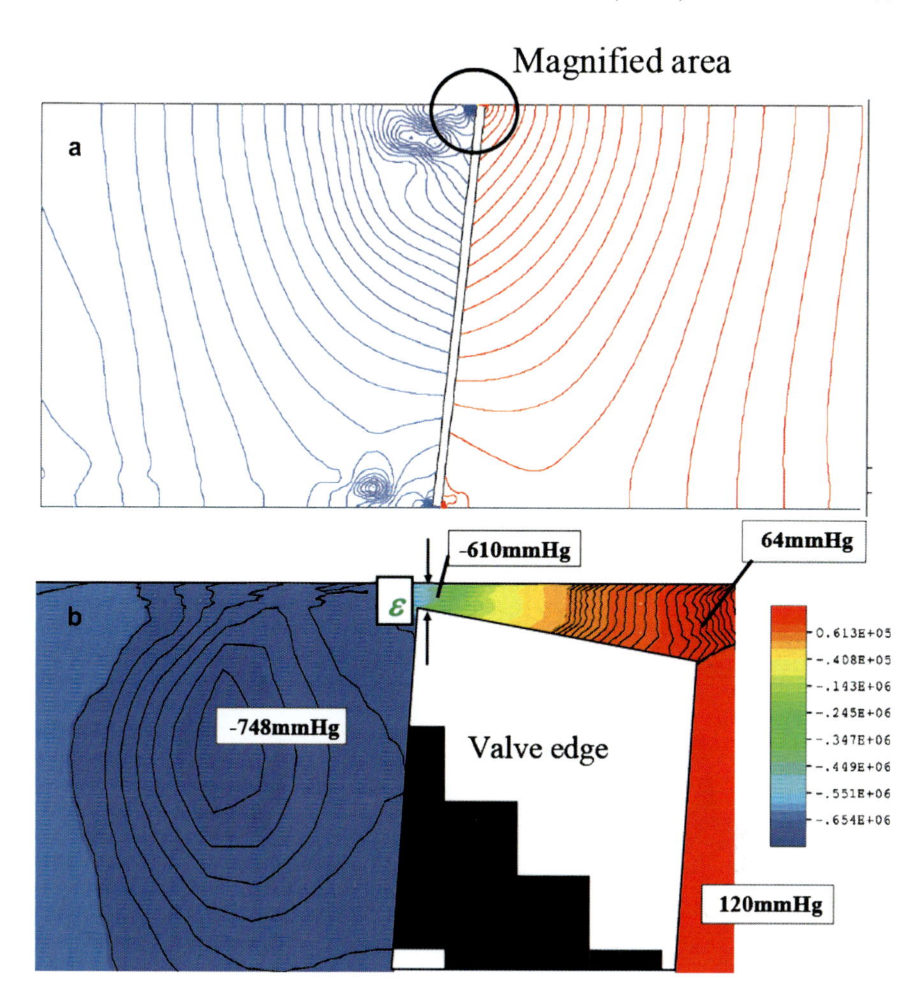

Fig. 4.5 Backflow jet induces pressure drop during valve closure as described by a numerical model. (**a**) Patterns of streamlines in the vicinity of a MMHV tip at onset of leaflet closure; (**b**) pressure field in a magnified region near the upper tips of the valve, indicating a significant low pressure in the core of the upstream tip vortex (The images are adapted from Avrahami et al. (2000)

position, the valve orientation may significantly influence coronary blood flow and the level of turbulence (Kleine et al. 2002). It is recommended that the MMHVs being implanted such that the large orifice of the valve opens to the non-coronary sinus (right posterior aortic wall; Fig. 4.6). This orientation guarantees that the two non-symmetric vortices fill the sinus and the flow developed in the ascending aorta is eccentric, and thus more similar to the natural spiraling flow pattern (Kvitting et al. 2010).

For MMHVs at mitral position, the optimal orientation is still under debate. Some surgeons recommend an anterior orientation to avoid the risk of leaflet impingements. However, most studies on LV flow advocate that the valve

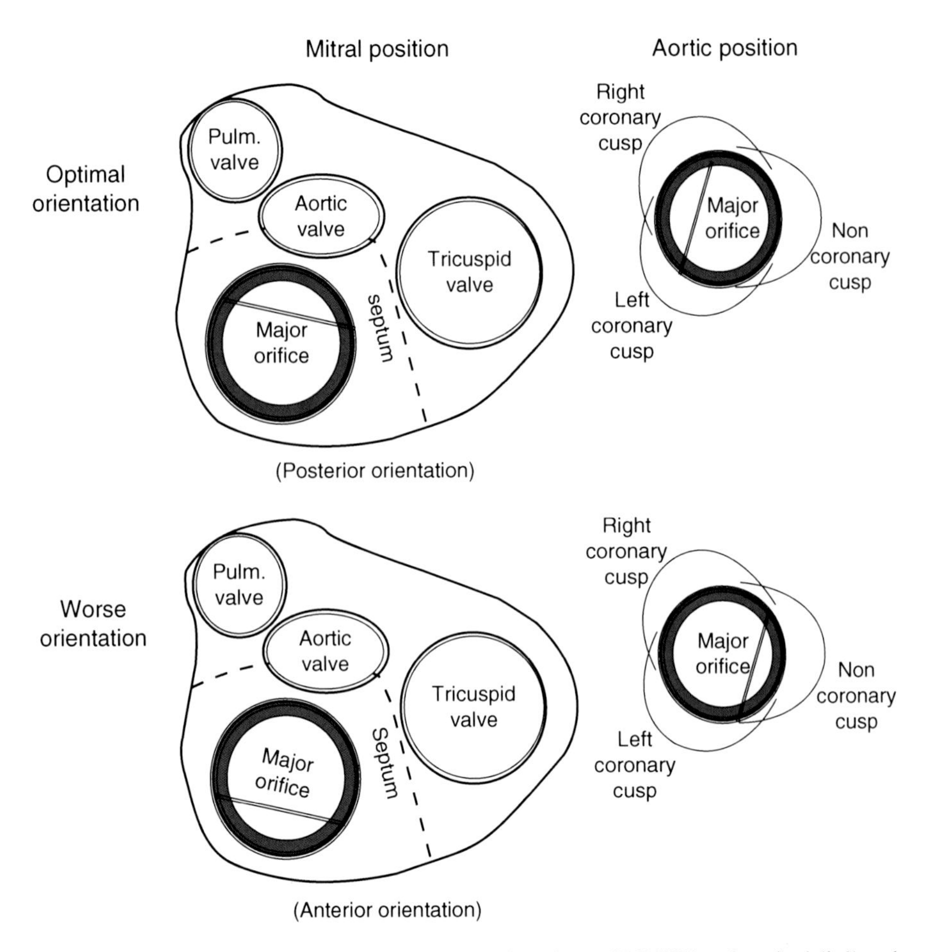

Fig. 4.6 Preferable (*top*) and unsuitable (*bottom*) orientations of MMHV at the mitral (*left*) and aortic (*right*) positions

implantation with the large orifice opening in a posterior fashion (Figs. 4.7 and 4.8), exhibits best hemodynamics due to producing stable circulation similar to the natural LV flow (Maire et al. 1994; Pop et al. 1989; Akutsu and Higuchi 2000; Machler et al. 2007). The flow formed in the LV by an MMHV implanted at the posterior is characterized by a large counterclockwise rotating vortex directed along the posterior wall towards the apex and back towards the aortic valve along the interventricular septum (Fig. 4.7). This transmitral vortex improves blood transport, and minimizes the loss of momentum when blood proceeds towards the aortic valve during systole (Mouret et al. 2005; Knapp and Bertrand 2005).

During the rapid filling phase of diastole, the strong shear layer generated by the jet emerging from the major orifice rolls-up and forms a vortex (Fig. 4.8). This vortex grows in size adjacent to the posterior wall, and proceeds towards the apex to fill

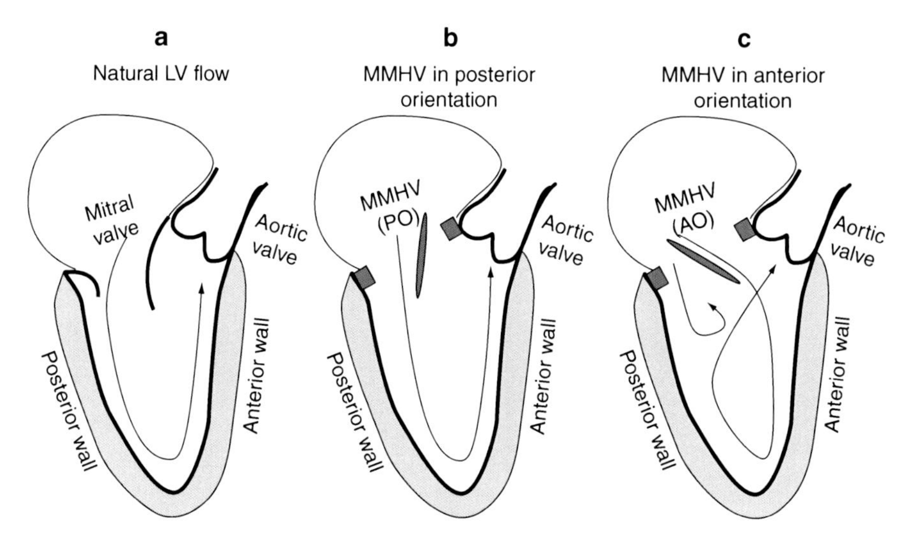

Fig. 4.7 (**a**) Direction of blood flow in the natural left ventricle compared to (**b**) the case with an MMHV implanted at the posterior orientation and (**c**) at the anterior orientation

the entire ventricle. The large vortex affects the flow in vicinity of the MHV, and the direction of the vortex shedding. The shedding vortices are swept with the leading vortex towards the apex where they are combined with it. The shear layer generated by the small jet emerges through the minor orifice and rolls-up into a steady vortex adjacent to the leaflet shielded by the wall during entire filling phases.

During atrial contraction, a second jet emerges through both orifices. At this phase, the inertia of the main vortex prevents the development of the small vortex downstream the minor orifice (Fig. 4.8). At the end of diastole, the increased pressure inside the ventricle along with the vortex induced rotational flow push the disk to close. During systole, direction of the flow is naturally toward the aortic valve while the predominant vortex core is pushed below the mitral valve where it is vanished and washed out of the ventricle. The regurgitant flow through the closed mitral valve generates local vorticity at the atrial side of the valve (Schoephoerster and Chandran 1991; Mouret et al. 2005; Knapp and Bertrand 2005; Rosenfeld et al. 1999).

For MMHVs implanted in an anterior orientation, the jets developed on the sides of the opened disk lead to formation of two separate vortices (Fig. 4.7). The majority of the inflow is directed through the greater orifice in a central direction toward the septum, and then reflected back by the septum to form a large reverse rotating vortex filling the basal two-thirds of the ventricle. An additional small vortex is generated from the smaller orifice. The two vortices interact with each other, and with the walls resulting in energy dissipation due to viscosity and turbulent perturbations. During early systole, the ejected blood through the aortic outlet interferes with this non-anatomical vortex, resulting in further turbulent stress, energy dissipation, and pressure drop.

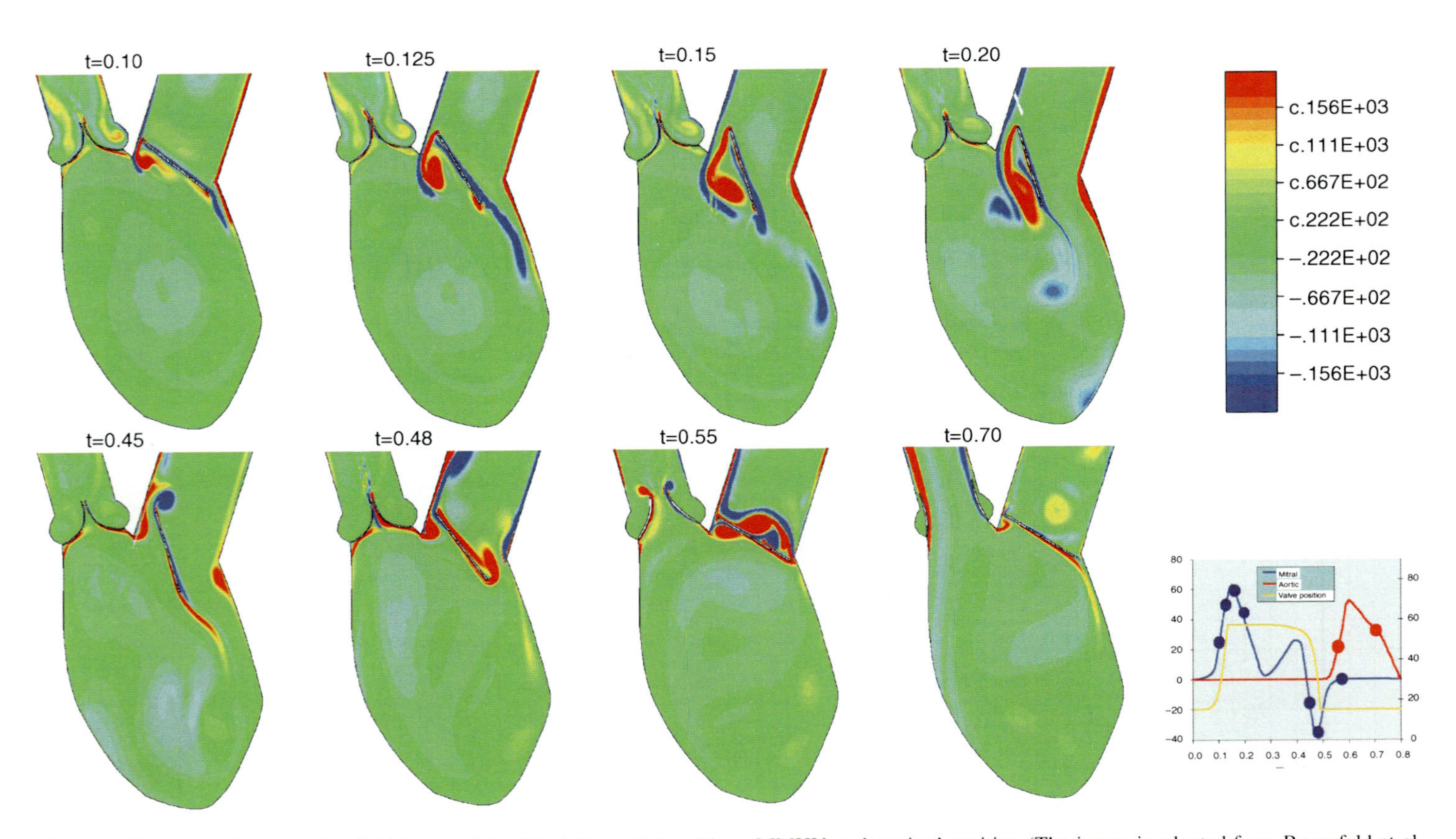

Fig. 4.8 Time-dependent vorticity field in a model of the left ventricle with an MMHV at the mitral position (The image is adapted from Rosenfeld et al. (1999))

4.2.5 Flow Across Bileaflet MHV (BMHV)

The current MHVs' worldwide market is dominated by bileaflet MHVs (BMHVs) with nearly 45% of all failing native valves replacements. However, despite their widespread use, BMHVs are not complication-free and are still associated with high levels of hemolysis, platelet activation and thromboembolic events. Unlike the flow across MMHVs, the flow across BMHVs is characterized with a symmetric central flow. While open, the two semicircular leaflets divide the valve's orifice into three regions: two semicircular lateral orifices and a narrow central rectangular orifice, forming a triple-jet flow structure. The major component of the forward flow emerges from the two lateral orifices. The pressure drop across BMHVs is somehow greater than MMHVs' (Nobili et al. 2008; Dumont et al. 2007).

During the opening phase, the leaflets of BMHV swing open completely, parallel to the direction of the blood flow, thus providing the closest approximation to central flow. As the valve opens and the flow accelerates, four shear layers are developed by the jets emerging through the orifices; one from each side of the leaflets. These shear layers roll-up and shed from the valve's leaflet edge, producing von Karman-like eddies (see Sect. 2.4 and Fig. 2.5 therein) at the wake of the leaflets (Kaminsky et al. 2007; Borazjani et al. 2010). The vorticity magnitude increases around and immediately distal to the valve leaflets, near the valve ring among the three jets. At these regions, high turbulent shear stresses are found. As the transmitral jet travels further downstream the valve, it becomes even more disturbed.

During the deceleration phase, the jets downstream the valve transform to a highly-chaotic turbulent-like flow, almost entirely dominated by small scale eddies and complex vortical interactions. These phenomena are typical for intermittent turbulent flows in the transitional range. Once the pressure gradient across the valve is inverted, the leaflets start to close passively. Similar to MMHVs, the leaflets' closing mechanism is thought to cause various valve conditions such as squeeze leakage flow, water hammer effect, elevated shear stress and extreme pressure gradients. The local fluid dynamics in the gap width during valve closure is characterized by large flow velocities, high shear stresses and highly vortical flow in the upstream side of the occluder (Fig. 4.9) (Krishnan et al. 2006; Cheng et al. 2004). The leaflets' dynamics during the closing phase induces large negative pressure values, which promote cavitation and microbubbles inception (Maines and Brennen 2005; Aluri and Chandran 2001). This may result in high potentials for formation of thrombi and microbubble emboli. The cavitation potential due to BMHVs is less than that of MMHVs. Most of the cavitation bubbles are found at the leaflets' tips, and the cavitation intensity correlates with the tip closing velocity (Lee et al. 2007; Herbertson et al. 2008). The maximal leaflet tip closing velocity for BMHVs is in the range of 2.4–3.2 m/s. The maximum velocity of the leakage flow in the closing phase falls within the range of 3.5–4.4 m/s with impact forces between the leaflets and the housing in range of 80–140 N (Mohammadi et al. 2006).

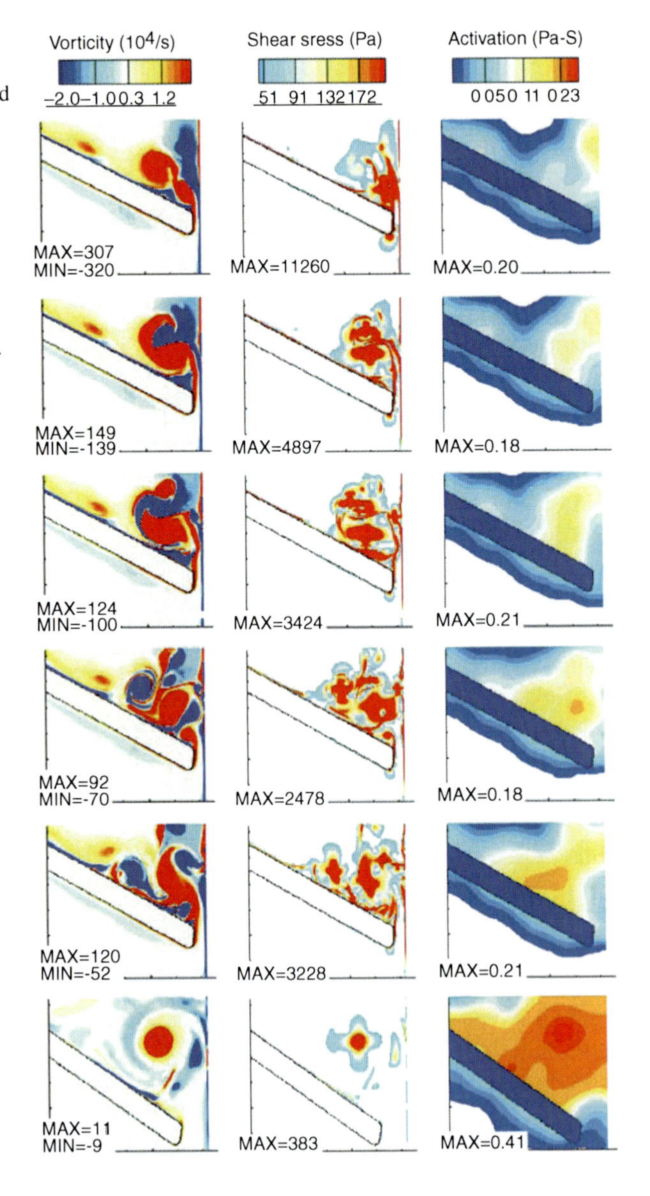

Fig. 4.9 Vorticity, shear stress and simulated platelet activation plots at closure and rebound stages of a BMHV obtained through numerical analyses. The *first panel* in each figure shows vorticity, the second, shows absolute shear stress and the third shows the activation parameter. The position of the leaflet indicated in (*f*) is 5 ms after the first impact of the valve with the housing. The strong vortices are advected away from the leaflet and diffuse over time (The images are adapted from Krishnan et al. 2006)

When the leaflets are fully closed, the BMHV leaflets are not totally sealed, leaving small gaps among the closed leaflets, the housing and in the hinge region. These gaps in BMHVs allow regurgitant flow, which is slightly more than that of experienced by MMHVs (Kaminsky et al. 2007). The regurgitant flow produces vortical flow structures upstream the leaflets (Cheng et al. 2004), where blood may be trapped due to a relatively large residence time, providing an ideal condition for platelet aggregation and formation of thromboemboli.

4.2.6 Consequence of Fluid Dynamics Experiments on Design of BMHVs

Clinical reports and recent *in vitro* experiments suggest that the flow at the hinge regions is a predominant factor in flow-induced complications of BMHVs. The hinge region was originally designed to prevent thrombus formation due to blood stasis, and allowing some regurgitant flow during the valve closure. However, the regurgitant high velocities through the hinge regions were found to induce higher shear stress and stable recirculatory flow upstream the hinge. The combination of high-shear stress and long residence time allows platelet to be activated, aggregated and deposited around the hinge region. Different pattern of clot formation were found in different designs of BMHV. Indeed, due to the major role of local hinge features on the global hemodynamic characteristics of BMHV, the specific design of the BMHV has a significant effect on its thromboembolic potential. Therefore, careful attention has been given to the complex flow dynamics around the recessed hinge and in the hinge pocket (Simon et al. 2004; Travis et al. 2001; Shu et al. 2004; Kelly 2002; Simon et al. 2010; Medart et al. 2004; Govindarajan et al. 2009b; Saxena et al. 2003; Wang et al. 2001; Ellis et al. 2000; Gross et al. 1996).

The forward flow in the hinge pocket region is highly complex and three-dimensional, with areas of recirculation, stagnation and flow reversal. During back-flow, two strong leakage jets are accompanied by high-velocity reversal flow at the hinge (Simon et al. 2010). The highest magnitude of shear stress was found at the hinge regions during the late stages of valve closure (Govindarajan et al. 2009a). As mentioned earlier, the hemodynamic parameters may significantly vary with valve design. For example, peak leakage velocity and peak turbulent shear stress measured at the hinge regions of different BMHVs may vary between 0.75 and 4.00 m/s and 2,000–8,000 dyn/cm^2, respectively (Yoganathan et al. 2004). Nevertheless, it has been shown that the thromboembolic potential of backwards leakage flow during the closed valve phase is larger than the forward flow in all BMHV designs (Dumont et al. 2007).

In addition to hinge geometry, some investigators also examined the effect of leaflet curvature on the forward flow across the BMHVs (Lee et al. 2009; Grigioni et al. 2001; Bang et al. 2005). They concluded that curved leaflet allows larger central flow, which reduces the velocity and turbulence. Another issue related to BMHV design is the asymmetric behavior of the leaflet, which results in asymmetry of the leakage jet distribution (Travis et al. 2002), or even valve dysfunction (Smadi et al. 2010). It is shown that asymmetric leaflets may alter the flow field downstream the valve, and the wake generation from the leaflet's trailing edge. Malfunction of a leaflet (e.g. due to tissue overgrowth or thrombus formation) limits its smooth motion, and severely disturb the flow-field by creating aberrantly high velocities and shear stresses, as well as large scale vortices.

4.2.7 BMHVs at Aortic Position

Similar to MMHVs, position and orientation of BMHVs critically affect the flow field passing through it. At the aortic position, three sinuses of Valsalva in the aortic root and the anatomic geometry of the aortic arch force an asymmetric and complex 3D flow field that depends on valve orientation (Nobili et al. 2008; Borazjani et al. 2010). It is recommended that BMHVs are implanted in such a way that one leaflet is directed toward the right coronary ostium (Laas et al. 1999) (Fig. 4.10). This type of implantation results in less turbulent flow and higher coronary perfusion (Kleine et al. 2002).

The main effect of the aortic root is the formation of recirculatory flow in the sinus regions. Upon implantation, the natural anatomy of the sinus is significantly modified due to the aortotomy itself, and implantation of the sutured ring in the cavity of the sinus. The shear layers developed by the side-jets roll-up into a ring-like structure (Fig. 4.11). That vortical structure expands towards the modified sinus cavity, and

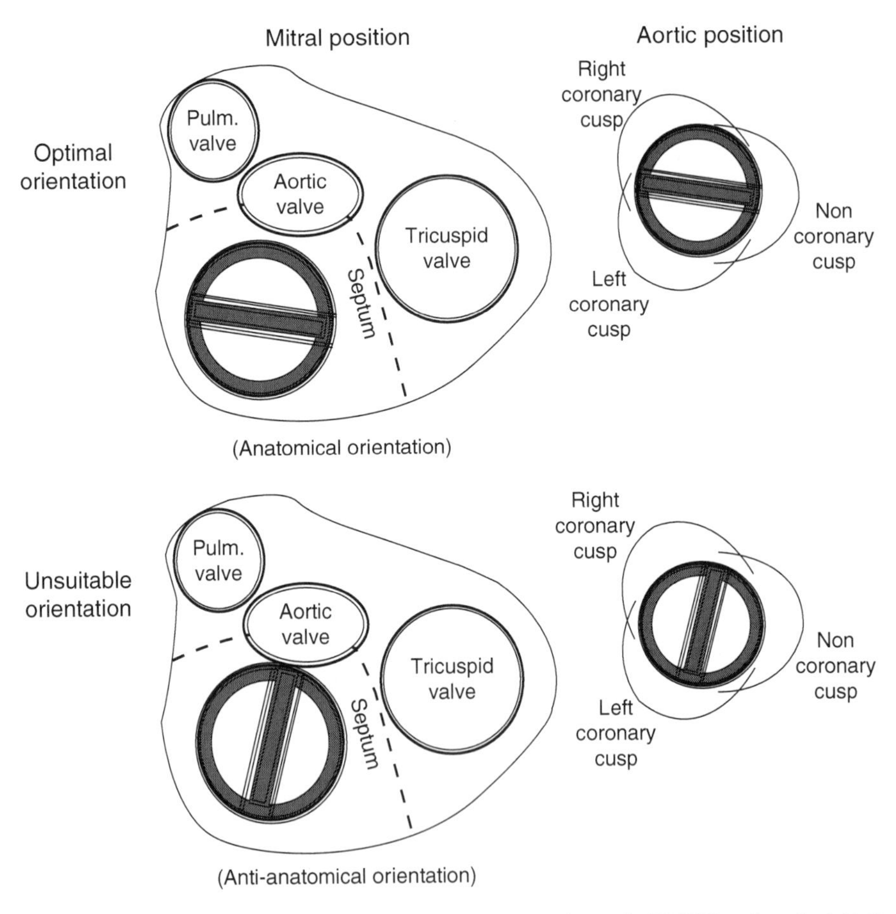

Fig. 4.10 (*Top*) Preferable, and (*Bottom*) unsuitable orientations of a BMHV at the mitral (*left*) and aortic (*right*) positions

stays there during most of the acceleration phase, taking the shape of the local sinus anatomy. At the onset of the deceleration phase, these sinus vortices become unstable, and break down into small chaotic structures with elevated turbulent dissipation (Li et al. 2010) (Fig. 4.11f–h). In addition to the local expansion at the sinus, the anatomic curvature of the aorta results in a significantly asymmetric flow condition. Consequently, the forward jets are suppressed and defuse rapidly downstream the aortic root, leaving behind local turbulence and recirculation downstream the wake of the leaflets. This asymmetry can be further protruded due to either valve misalignment or oversizing (Bluestein et al. 2002; Chambers et al. 2003).

Another important feature of the aortic anatomy is the effect of sinus vortices on leaflet closure. As mentioned earlier, at the end of systole, the vortical flow inside the sinus dramatically changes. In case of the natural aortic valve, this change in the vortex' rotational direction induces an inward motion of the leaflets that pushes the leaflets to close. However, for the BMHVs, since the closing dynamics requires outwards leaflets motion, the vortical flow in the sinus prevents them from closing. Thus, the leaflets are closed only in a later stage due to the negative pressure across the valve, sweeping larger volume of regurgitation (Ohta et al. 2000). Additionally, the consequent rapid slam of the leaflet closure against the flow direction causes water-hammer effect, leaflet rebound, cavitation and fracture of the leaflets (Choi and Kim 2009).

4.2.8 BMHVs at Mitral Position

The flow across BMHVs at the mitral position is strongly affected by the anatomic geometry of the LV and the orientations of the valve. Surgeons often tend to position a mitral BMHV at the anatomic position, with the central orifice perpendicular to the septum (Fig. 4.10) to mimic the natural valve shape. However, this orientation for the valve was found to be less favorable due to late closure of the posterior leaflet, and potential risk of leaflet impingement by the posterior wall and the posterior leaflet thrombosis. The preferred performance of BMHVs in anti-anatomic orientation, where the tilting axes are parallel to the ventricle septum (Fig. 4.10) is clinically shown (Van Rijk-Zwikker et al. 1996; Schoephoerster and Chandran 1991; Akutsu and Higuchi 2000). From hemodynamics perspective, the LV flow is practically similar regardless of the valve's orientation. The anti-anatomic orientation offers a slightly better hemodynamics, with reduced turbulence and smoother washout near the LV free wall (Fig. 4.12) (Akutsu and Higuchi 2000). Conversely, the anti-anatomic orientation is attributed to some increased flow disturbances near the apex mostly due to the spiral ejection in the LV (Machler et al. 2004).

Nevertheless, both orientations result in flow patterns that significantly deviate from those in the natural heart (Fig. 4.12). The three central jets are directed towards the septum and then reflected back by the ventricular septum to form a penetrating vortex ring, which results in complex vortical components filling the ventricle at peak diastole (Faludi et al. 2010; Kheradvar et al. 2010). This vortical flow structure

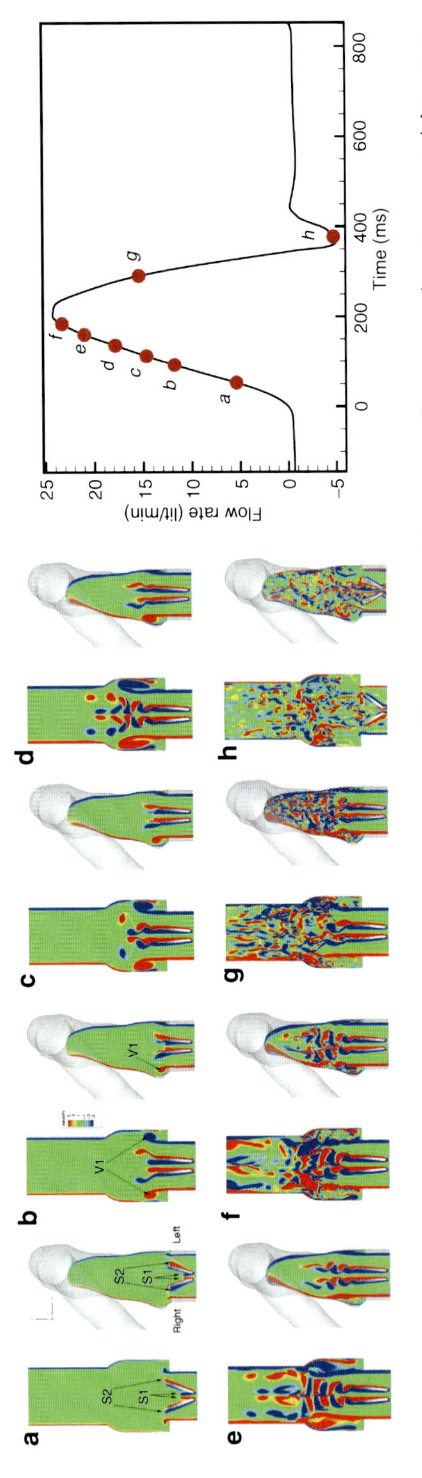

Fig. 4.11 Flow across BMHV at the aortic position: Numerical comparison of instantaneous vorticity contours for an anatomic aorta vs. a straight aorta on the mid-plane of BMHV (The images are adapted from Borazjani et al. (2010))

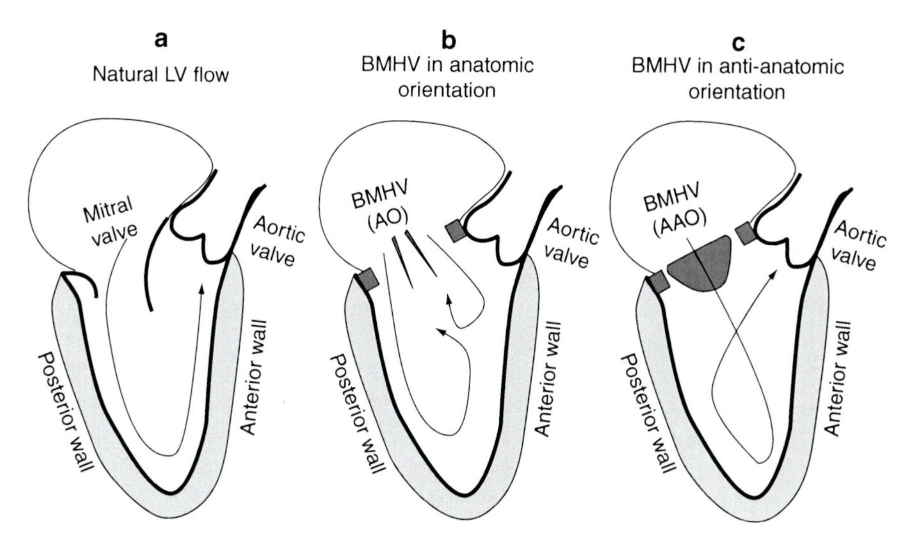

Fig. 4.12 (**a**) Flow direction in the natural left ventricle, and in presence of a BMHV at (**b**) the anatomic orientation and at (**c**) the anti-anatomic orientation

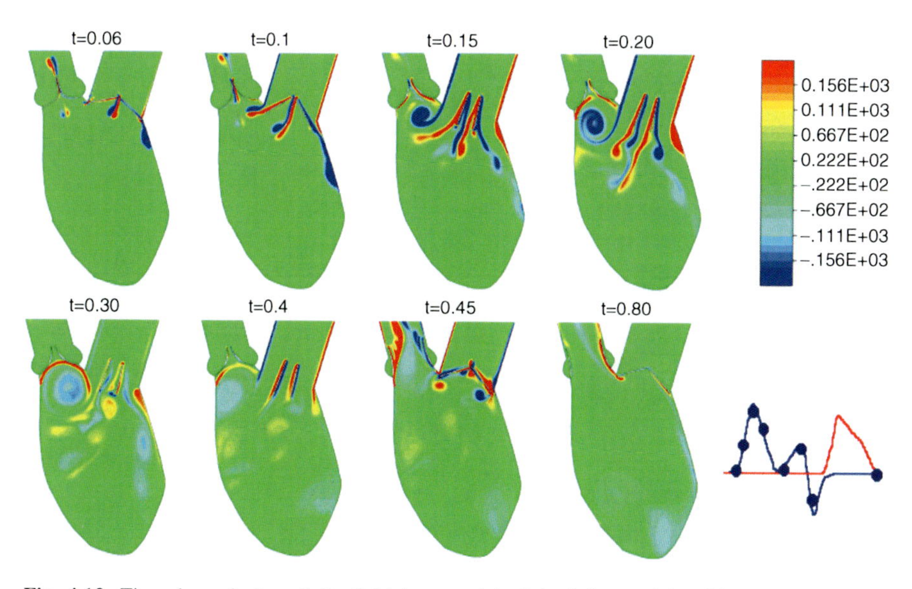

Fig. 4.13 Time-dependent vorticity field in a model of the left ventricle with a BMHV at the mitral position (The image is adapted from Avrahami et al. (2002))

is different from the large rotating vortex generated in the natural ventricle during diastole (Fig. 4.12). The non-anatomical flow regime produced by the BMHVs' central flow and the wake generated by the side-jets yield significant energy dissipation through viscosity and turbulent perturbations in the LV during diastole (Pedrizzetti et al. 2010). In addition, the anterior orifice jet, rolls-up into a small vortex below the aortic valve (Querzoli et al. 2010; Avrahami et al. 2002), where the local low pressure may increase aortic valve regurgitation (Fig. 4.13).

4.3 Vortical Flow Structures in Ventricular Assist Devices (VADs)

Ventricular Assist Devices (VADs) are mechanical circulatory devices that are used to support or replace the function of a failing ventricle. VADs offer a life-saving alternative for patients with end-stage heart failure who are unqualified or unable to wait for heart transplantation. By assisting the function of the damaged ventricle, VADs help restoring normal hemodynamics and supporting the cardiac function either as a bridge to heart transplantation/myocardial recovery or as the destination therapy. With the increasing shortage in organ donation and the proven superiority of VADs over drug therapy (Rose et al. 2001), the need for VADs as a long-term chronic support becomes essential (Clegg et al. 2006). Therefore, great efforts are being made for improving VADs design, implantation techniques and patients managements (Birks 2010). Yet, the VADs are far from being complications free. The survival rate among patients with first-generation VADs in the REMATCH study (Rose et al. 2001) during first and second years were only 52% and 23%, respectively (Fig. 4.14).

Large experience acquired over the past 20 years with the end-stage patients contributed to optimization of therapy management, and thus resulted in improved patients' survival (Osaki et al. 2008; Cooper et al. 2010; Holman et al. 2010). Previous studies report survival rate of 58% after 1 year for patients who received LVAD as destination therapy (Lietz et al. 2007; Fang 2009). Additionally, more recent studies suggest that latest VAD designs offer improved results, with survival rates at first and second years of up to 68% and 58%, respectively, as shown in Fig. 4.14 (Daneshmand et al. 2010; Lahpor 2009; Saito et al. 2010; Pagani et al. 2009; Slaughter et al. 2009).

According to clinical studies, 30–55% of VAD complications are related to device malfunction and driving line infection (Rose et al. 2001; Minami et al. 2000; Martin et al. 2006; Park et al. 2005). The major mechanical requirements for such devices are reliability, autonomy, and energy efficiency. The ultimate goal of VAD research is to develop a highly efficient, fully-implantable, small and light pump for long-term use outside the hospital setting. The ultimate device should meet the challenge of sufficient cardiac output with 40 million beats per year. Other preferences are durability, minimal number of moving parts, biocompatibility and low cost (Clegg et al. 2006; Daneshmand et al. 2010; Saito et al. 2010; Okamoto et al. 2006).

Improving VADs' hemodynamics is also of great importance. Thrombus formation, fatal hemorrhage and stroke are common life-threatening complications. The reported thromboembolic complications developed due to VAD support is around 47% (Minami et al. 2000; Rothenburger et al. 2002). These complications are partially linked with fluid dynamics of the artificial heart technology, such as hemolysis caused by excessive turbulence and clot formation incurred by regions of flow stagnation. Most of these complications are addressed through the new designs,

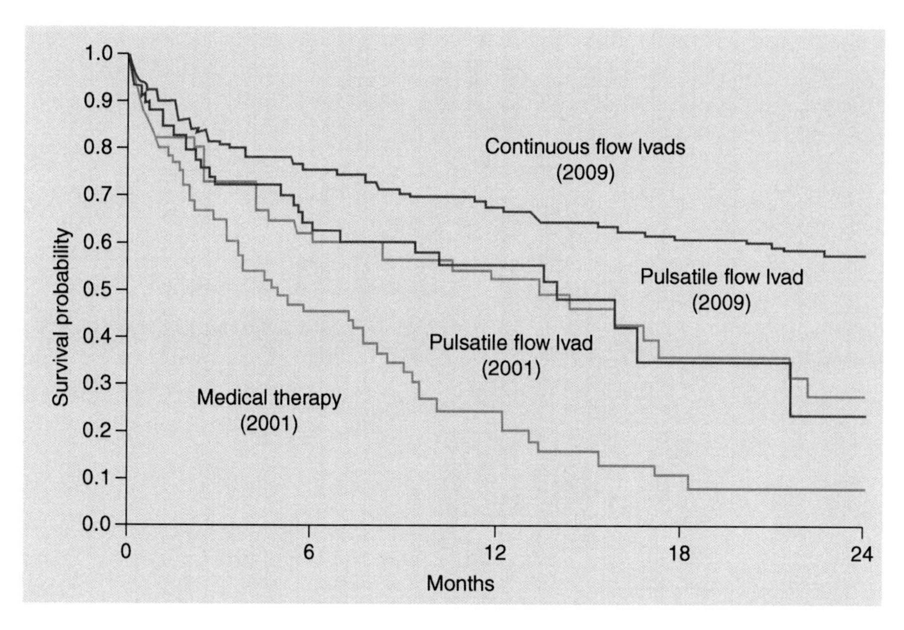

Fig. 4.14 Survival rates in two trials of Left Ventricular Assist Devices (LVADs) as destination therapy. The curves labeled 2009 are those reported by Slaughter et al. (2009); those labeled 2001 were reported for the REMATCH trial (Rose et al. 2001) (The image is modified from (Fang 2009))

however, other complications such as gastrointestinal bleeding are yet to be resolved (Miller 2010).

4.3.1 Classification of VADs

Two main categories of long-term implantable VADs pumps are: *pulsatile pumps*, that mimic the natural pulsing action of the heart, and *continuous flow pumps*. Pulsatile VADs (e.g. HeartMate® I, Novacor® LVAS) use positive displacement pumps, including a blood chamber with inlet/outlet conduits equipped with valves that direct the flow from the heart to the aorta. They are usually larger in size compared to continuous pumps, considered more complicated and consume more power.

Continuous flow VADs such as Jarvik-2000 and HeartMate® II involve rotary-pump technology to provide non-pulsatile blood flow. They use either centrifugal pumps or an axial flow pump. These pumps are smaller, have a low power consumption and considered more durable compared to pulsatile pumps (Slaughter et al. 2009; Mahmood et al. 2000). Current scientific evidence suggests that constant flow without pressure oscillations on a short-term basis is equivalent to native pulsating cardiovascular function (Allen et al. 1997). Whether this non-pulsatile flow is as effective as pulsatile flow for permanent use is still undetermined (Slaughter et al. 2009; Miller 2010; Mahmood et al. 2000; Sezai et al. 1999; Nose et al. 2000; Travis et al. 2007; John et al. 2009; Undar 2004).

Other types of artificial pumps include total artificial heart (TAH) pumps (Yang et al. 2007; Copeland et al. 2004), which include two ventricles and four valves, to replace the complete heart, and physiological cardiac assist devices (PCAD) (Landesberg et al. 2006; Naftali et al. 2006), which is a pulsatile VAD working in conjunction with the heart and is synchronized with the natural heart rhythm, supplying the necessary additional energy required for patient's normal function. The device is connected to the apex of the heart with only one cannula without a valve.

4.3.2 Pulsatile VAD Hemodynamics

Complications related to thrombus, emboli or fatal hemorrhages are closely linked to the fluid dynamics of the pump, including: interaction of blood with the moving components, high shear stresses, turbulence, and separated or stagnant flow regions (Minami et al. 2000; Mahmood et al. 2000; Affeld et al. 1997; Oley et al. 2005; May-Newman et al. 2006; Argueta-Morales et al. 2010). These complications vary drastically from one VAD type to another. Currently, the device-related thromboembolic complications with minimal or no anticoagulant therapy is reported to be around to 2–3% (Mahmood et al. 2000).

The areas that are mostly prone to thrombus formation in pneumatic pumps include the heart valves, connectors and the native ventricle (Mahmood et al. 2000; Deutsch et al. 2006). The flow through narrow conduits and valves is characterized as highly disturbed with high velocities and high shear stresses. The valves' hemodynamics are strongly related to their type, orientation and the conduits in which they are placed in (May-Newman et al. 2006; Affeld 1998; Yin et al. 2004; Avrahami et al. 2006a). The flow inside the chamber is characterized by lower velocities and shear stresses, with longer residence times. Therefore, the major contributors to the thrombus formation inside the chamber are blood mixing and washout properties. The combination of high shear stress regions where platelets are activated and the stagnation regions where platelets aggregate provide an ideal situation for thrombus formation (Bluestein et al. 2002). Figure 4.15 demonstrates the effect of VAD design and valves on its hemodynamics.

4.3.3 Valves in the VAD

Artificial heart valves that are often used to replace a diseased natural valve are also utilized in pulsatile VADs. As mentioned in previous sections, mechanical heart

	FM	FB	PPB
VAD chamber models illustrating chamber at maximal compression	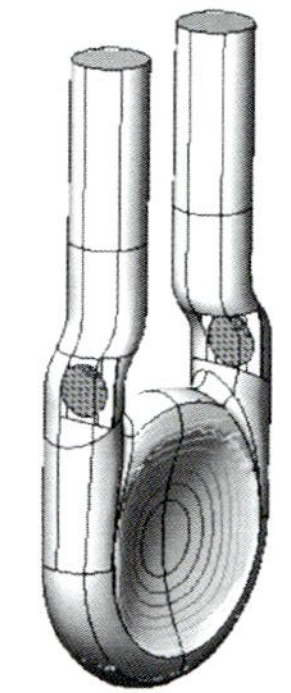	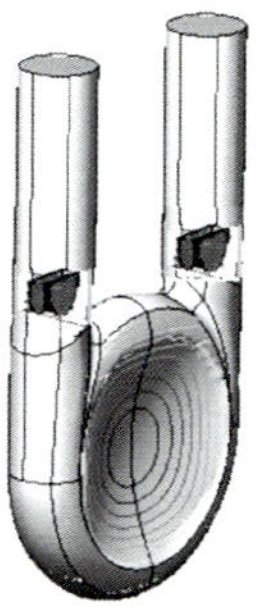	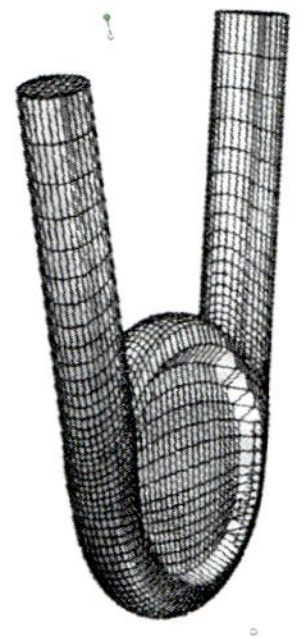
black line illustrates the motion of the center of the major vortex during filling	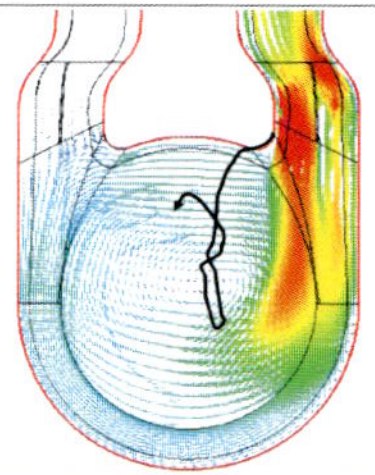	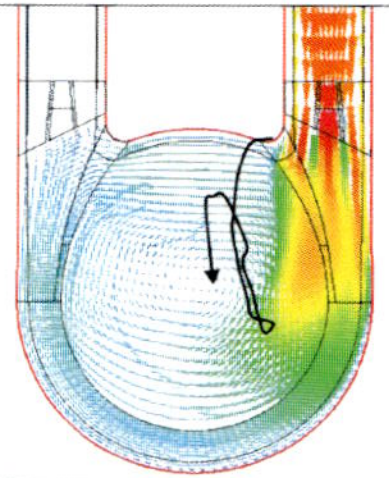	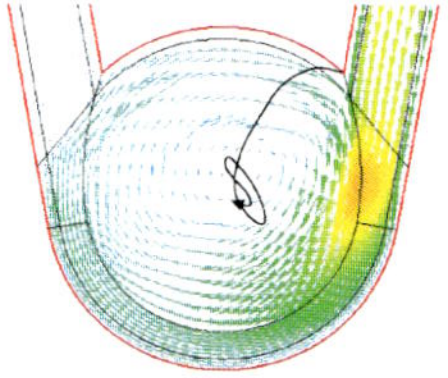
Maximal residence time	6 cycles	4 cycle	3 cycle
Max shear stress	6 Pa	9.5 Pa	3 Pa
Maximal stagnation period	0.33sec	0.3sec	1.5sec

Fig. 4.15 An example of the effect of valves and compression mechanism on chamber mixing and washout properties. The figure shows three different VAD chamber designs as follows: *FM* Flexible chamber with Monoleaflet valves (S-shape conduit), *FB* Flexible chamber with bileaflet valves (straight conduit), *PPB* Chamber with dual flat pusher-plates and bioprosthetic valves. The motion of the center of the vortex during a pump cycle is described using *an arrow*. Shear stress, washout and stagnation parameters are listed (Avrahami 2003)

valves are associated with high risk of thromboembolism and stroke due to the development of highly disturbed flow patterns. Since each VAD contains two valves (inlet and outlet), the valves play critical roles in the VAD's hemodynamics, as they direct the flow in and out of the chamber. Moreover, the largest magnitudes of wall shear stresses and bulk turbulence stresses in VADs occur in the vicinity of the inflow and outflow valves (Baldwin et al. 1994; Mussivand et al. 1999).

The effect of valve design has been investigated in several studies where even small modifications in the valve's leaflets (Lee et al. 2009) or orientation (Kreider et al. 2006) made a dramatic impact of the VAD hemodynamics performance. Therefore, investigators tried to develop VAD-specific valves to optimize its performance (Escobedo et al. 2005; Iwasaki et al. 2003). In a VAD system, it is possible to orient an MMHV in an S-shape conduit, thus the leading edge of the occluder would be in line with the incoming flow. In this configuration, the flow separation is minimal and the resistance to the flow would be lower. Consequently, the pressure drop across the valve is only 60% of conventional MMHVs (Affeld 1998). Indeed, the mixing and washout properties exhibited by the MMHV configurations are relatively poor. Yet, no lasting stagnation regions are found and the overall cumulative risk for thrombus formation is smaller than in the BMHVs. Therefore, MMHVs may be considered more favorable to be used in VADs (Avrahami et al. 2006a; Avrahami 2003).

4.3.4 VAD Chamber

The blood chamber's geometry and volume-changing mechanism are critical for blood hemodynamics, mixing and washout inside the VADs. Improvements in chamber's design and compression mechanism are limited by the mechanical considerations. Even small design modifications may have a potential to dramatically affect the VAD's hemodynamics (Oley et al. 2005; Mussivand et al. 1999; Bachmann et al. 2000; Rose et al. 2000; Okamoto et al. 2003; Slater et al. 1996; Goldstein et al. 1998; Avrahami et al. 2006b; Hochareon et al. 2004a; Doyle et al. 2008; Moosavi et al. 2009). For example, small projections (of 4 mm in height) placed on the chamber's housing or diaphragm have resulted in major variation of the flow inside the chamber (Sato et al. 2009). Alternatively, minor geometric changes of the chamber has shown to reduce the peak shear stress (Mussivand et al. 1999).

Blood chamber's washout is an important parameter in design of the VADs. Several pump cycles usually take for all the blood particles to be washed out from the chamber. Chambers with longer residence time may induce higher risk for platelets aggregation and thrombi formation. One way to evaluate blood mixing and residence time in the chamber is to inject particles or dye into the chamber during a filling phase, and track them during several pump cycles. (Goubergrits et al. 2008; Naftali et al. 2006; Avrahami et al. 2006a; Medvitz et al. 2009; Koenig and Clark 2001). One of the major flow features inside a VAD chamber is the incoming shear layers followed by a large rotating vortex during filling phase. The stagnant flow in the center of this vortex may induce platelet aggregation. If the

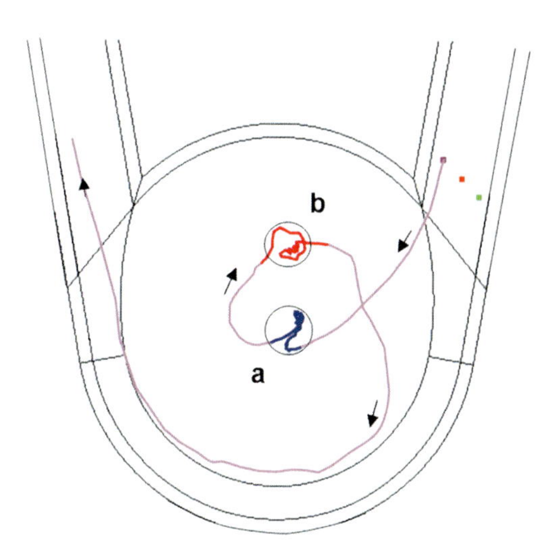

Fig. 4.16 Particle tracking during three pump cycles in a model of a flat pusher-plate chamber. The particle is trapped in the center of the steady vortex for a complete cycle (**a**) and then for another half a cycle (**b**), before it is released from the chamber (The image is adapted from Avrahami (2003))

vortex is steady and stable-as shown for the VADs with flat pusher-plates (Figs. 4.16 and 4.17), the risk of thrombi formation increases due to poor blood mixing and washout. Thus, the behavior of the major filling vortex is important.

Pump designs with hemispherical or flexible chambers have preferable washout characteristics (Rose et al. 2000; Slater et al. 1996; Goldstein et al. 1998; Avrahami et al. 2006b; Hochareon et al. 2004a). Flexible chambers are found to be advantageous over pusher plate configurations due to lower wall shear stress and minimal stagnant flow (Yang et al. 2007; Avrahami 2003; Christensen et al. 1993). However, poor membrane motion may reduce the pump's stroke volume and even increase the potential for thrombus formation (Hochareon et al. 2004a). Hemispherical chamber geometry was shown to encourage "traveling" of the main vortex during the filling phase, entraining fluid from stagnant regions in its path and improving VAD's hemodynamics. Similar "wandering vortices" in VAD chambers were observed *in vivo* using MRI (Markl et al. 2007), *in vitro* using PIV (Cooper et al. 2010; Avrahami 2003; Sato et al. 2009; Hochareon et al. 2004b) and computationally (Avrahami 2003; Doyle et al. 2008; Moosavi et al. 2009; Medvitz et al. 2009). The reduction of device-related thromboembolic complications in advanced HeartMate® LVAD is partly attributed to its hemispherical chamber (Slater et al. 1996; Goldstein et al. 1998).

4.3.5 Continuous Flow VADs (CF VADs)

Continuous-flow pumps are considered the second generation of VADs. They are smaller, lighter, noiseless, valve-free, more durable and have smaller theoretical thrombogenic foreign surface (Fang 2009). More recent studies show improvement in the rate of survival, quality of life, functional capacity of patients, and device durability with the CF VADs as compared with the pulsatile-flow VADs (Slaughter et al. 2009). However, CF VADs are also not without complications. The main complication of CF VADs is stroke (12%), which is similar to pulsatile-flow VADs (Argueta-Morales et al. 2010; Behbahani et al. 2009). Additionally, the interaction among blood, rotating blades and stagnant areas adjacent to the contact surfaces results in some degrees of hemolysis and platelet activation (Kawahito et al. 2000; Fan et al. 2010; Garon and Farinas 2004; Song et al. 2010). Excessive hemolysis that leads to anemia, and the free circulating hemoglobin that is toxic to kidneys, may lead to multiple organs failure (Sakota et al. 2008). Thromboembolisms may lead to stroke, neurologic deficits, or even death. Thus, long-term anticoagulation therapy should be administered for patients with CF VADs. Bleeding complications are more prevalent with CF VADs than with the pulsatile devices; in the range of 17–63 events per 100 patient-year (Uriel et al. 2010).

Recent studies suggest that bleeding complications in CF VADs are not only a consequence of anticoagulation therapy, but also are attributed to the non-physiologic flow in these devices. It has been shown that non-pulsatile flow affects organ function and hormonal situation (Slaughter et al. 2009; Sezai et al. 1999; Undar 2004). Although the mean blood pressure is often elevated in patients with CF VADs, vascular pulsatility significantly diminishes. The integrity of the vascular endothelium is partly dependent on the stretch and distension created by pulsatile flow. Prolonged diminished pulsatility may lead to endothelial dysfunction, vascular stiffening and higher vascular impedance that lead to a higher ventricular workload and reduced myocardial perfusion (Travis et al. 2007). As a consequence, non-pulsatile flow affects end-organ function, increases the risk of thromboembolism and plaque deposition (Birks 2010), promotes the development of arteriovenous malformations (AVMs), increases incidence of gastrointestinal bleeding (Miller 2010; Uriel et al. 2010) and ventricular arrhythmia (John et al. 2009).

Another complication related to non-pulsatile VADs is the aortic valve fusion and incompetence. The permanent low cardiac function due to CF VADs circulatory support adversely modifies myocardial perfusion, cardiac electrophysiology and aortic valve biomechanics. Under CF VAD circulation, the transvalvular pressure is high, and the aortic valve remains closed throughout the entire cardiac cycle producing a VAD-related functional stenosis (May-Newman et al. 2010). Recent designs introduced the magnetically levitated centrifugal blood pump whose rotor is completely suspended using full magnetic levitation. The full magnetic levitation eliminates bearings wear within the pump and improves blood compatibility by allowing greater clearance around the rotor and semi-pulsatile flow patterns (Song et al. 2010; Pai et al. 2010)

4.3.6 Hemodynamics of VADs Cannulation

The VAD inflow cannulation either to the atrium or ventricular apex of the native heart induces major changes in ventricular flow dynamics. To avoid ventricular wall tension and ventricular dilatation, it has been suggested that the LV apex cannulation maybe more advantageous (Korakianitis and Shi 2007) for improving hemodynamic parameters, such as ventricular ejection fraction (EF), stroke work, and pump flow rates (Timms et al. 2010). To avoid major disruptions in aortic flow and possible thrombus formation owing to the lack of blood washout due to VAD's outlet cannula, use of proximal cannula insertion is recommended (May-Newman et al. 2006).

4.4 Vortex Formation due to Arterial Surgery and Anastomosis

The asymmetries of flow and formation of vortices may also result from unnatural events such as surgical interventions. In these cases, the vortices can negatively influence energy dissipation, and may be a sign of inefficient flow.

4.4.1 Fontan Procedure

Surgical repair of complex congenital heart defects often impose major anatomical reconstructive procedures and create new cardiovascular circuits and/or connections. Methods – that are applied *in vitro* and *in silico* – have been frequently used to study the fluid dynamics of these complex hearts. A well-known example is the *Fontan procedure*, which describes a group of surgical procedures used to improve the circulatory system of children with hypoplastic left heart syndrome having a single effective ventricle.

The *Fontan circulation* refers to an anatomical configuration where the single ventricle pumps blood returning from the lungs to the body, and the blood returning from the body travels to the lungs via direct blood vessel connections without a pumping chamber. The three currently used surgical reconstruction categories in the treatment of univentricular heart are: (*i*) the total cavopulmonary connection (TCPC) (de Leval et al. 1988), (*ii*) the hemi Fontan procedure (HFP) (Douglas et al. 1999) or the bi-directional cavopulmonary connection (BCPA) (Haller et al. 1966), and (*iii*) either modified Blalock-Taussig shunt (MBTS; systemic-to-pulmonary shunt) (de Leval et al. 1981) or RV-to-pulmonary shunt (Sano et al. 2003).

In TCPC technique, the superior and the inferior vena cava (IVC) are directly connected to the right pulmonary artery through an intra-atrial or extra cardiac tunnel. The HFP consists of a connection between the undivided superior vena cava (SVC) junction and the pulmonary arteries along with insertion of a temporary intra-atrial patch that prevents blood to reach the lungs through IVC. This patch is eventually removed at the final stage of the repair to complete the TCPC. The alternative is the BCPA that consists of the connection between the SVC and the right pulmonary artery. The MBTS or the Sano shunt are used as the first stage palliation in treatment of hypoplastic left heart syndrome. Through this technique, the pulmonary valve is used as (1) the systemic ventricular outlet and (2) for construction of a systemic- or RV-to-pulmonary artery shunt to provide pulmonary blood flow.

The surgical repair should supply distal organs and vessels with proper distribution of blood flow. Additionally, an ideal surgical repair may not cause blood flow abnormalities (e.g. vortices, flow separation and stagnation areas) that induce vascular

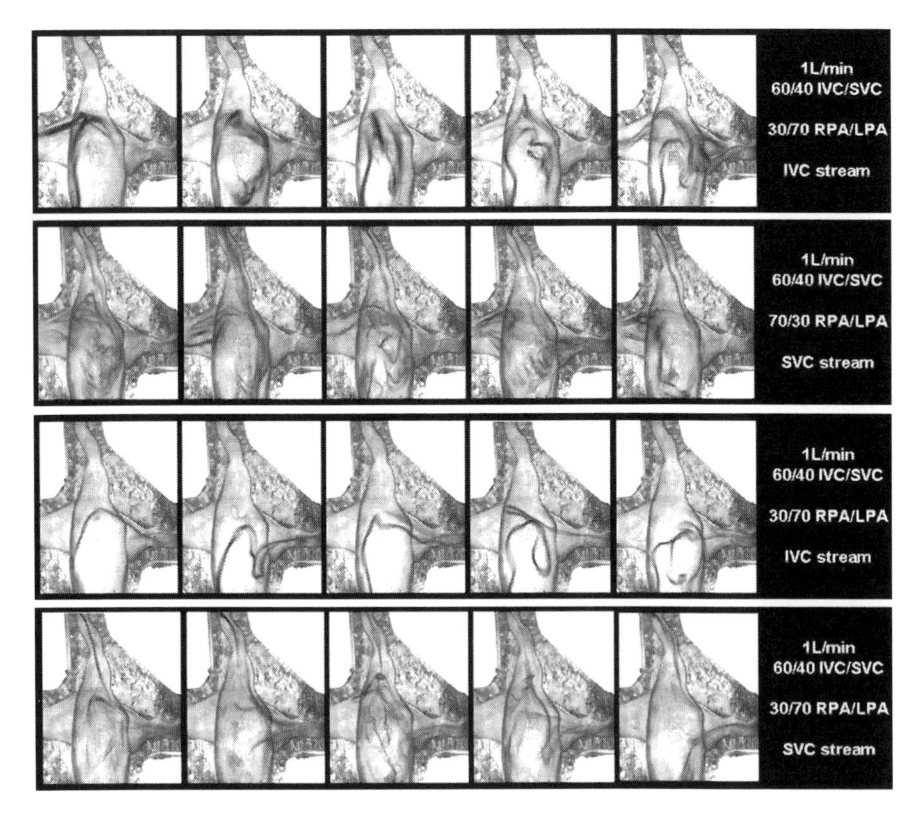

Fig. 4.17 Flow visualization of the intra-atrial model at 1 l/min (*IVC* inferior vena cava, *LPA* left pulmonary artery, *RPA* right pulmonary artery, *SVC* superior vena cava) (The images are adapted from de Zélicourt et al. (2005))

resistance, wall damage, thrombus formation or energy dissipation. Though, clinical studies have shown that most operational designs result in poor postoperative outcomes and significant energy loss. The clinical problem presented to the cardiac surgeon is design of an anastomosis with: (*i*) minimal energy loss, which requires avoiding even minimal gradients, areas of stagnation and 'right angles', (*ii*) potential for growth, and (*iii*) correction of blood distribution to lungs, coronaries and systemic organs.

The first computational models on cavopulmonary anastomoses (Van Haesdonck et al. 1995; Dubini et al. 1996; de Leval et al. 1996) utilized the finite element method. Van Haesdonck et al. compared the energy losses using a cavopulmonary compared to an atriopulmonary connection (Van Haesdonck et al. 1995). Dubini et al. and de Leval et al. suggested that the energetics can be improved by applying asymmetry between the SVC and IVC connections (Dubini et al. 1996; de Leval et al. 1996), which was later confirmed by other studies (Bove et al. 2003; Khunatorn et al. 2002; Migliavacca et al. 1999; Ryu et al. 2001).

de Zélicourt et al. performed a TCPC *in vitro* and showed that high power losses are present either in the absence of a caval offset or in the presence of large anasto-

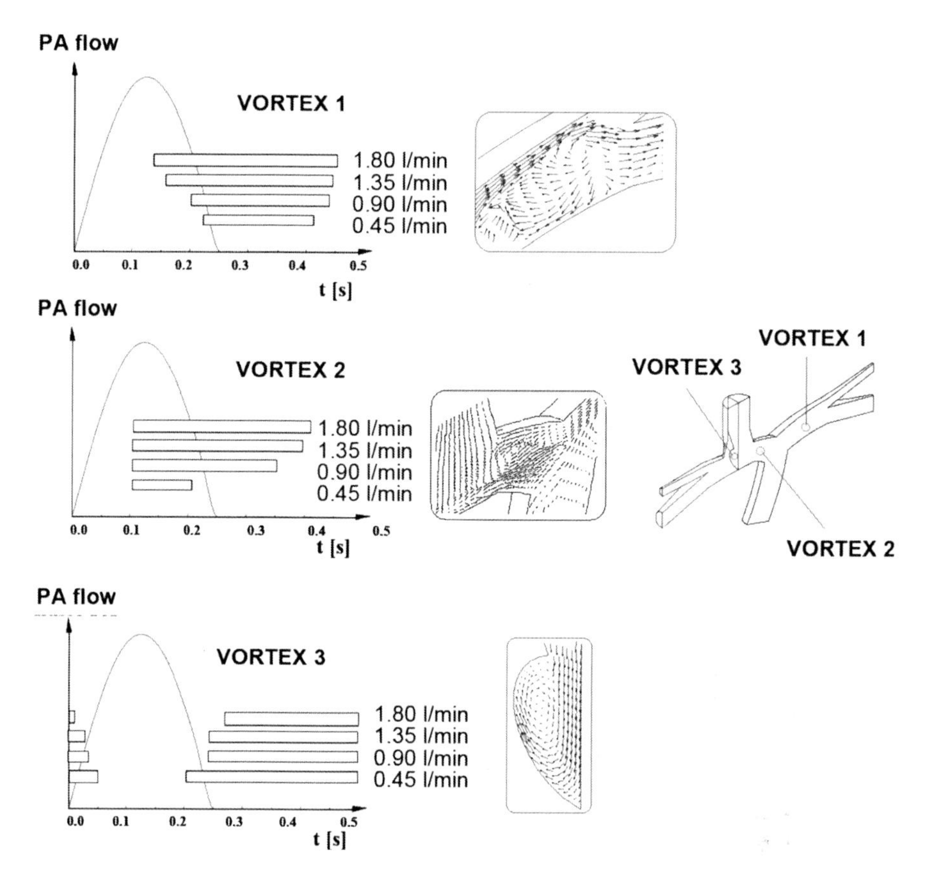

Fig. 4.18 Presence and deposition of the three main vortices during the cardiac cycle superimposed on the input blood flow volume waveform (*PA inlet*). All of them are present during diastolic period. Vortex 2 starts after the systolic peak; only in simulation with the lowest PA flow, the vortex is absent in diastole because of the flow from the SVC to the LPA. A selection of the corresponding velocity vector plots are reported in the insets (The images are modified from Migliavacca et al. (1996))

motic area where the diameter of pulmonary vessels are different than those in the connection area (de Zelicourt et al. 2005). Most of the energy was dissipated in the pulmonary arteries (PAs) through wall friction. Thus, the diameter of the PAs is a crucial factor, particularly in patients with PA stenosis. The connection area should match the diameter of the connecting vessels as closely as possible because dimension mismatch may enhance the vortex formation at the center of the connection. Additionally, the diameter mismatch may yield to higher energy dissipation, flow separation and stagnant flow regions (Fig. 4.17).

Studies based on the local effects of pulsatile forward-flow on the native pulmonary artery in the BCPA showed that the development and extension of vortices adjacent to the main pulmonary artery is correlated to the volume of blood coming from the RV (Fig. 4.18), which in turn is related to the efficiency of the lung perfusion (Migliavacca et al. 1996, 1997).

Computer models have also been used to study the hemodynamic performance of cavopulmonary connections after the Norwood procedure for hypoplastic left heart

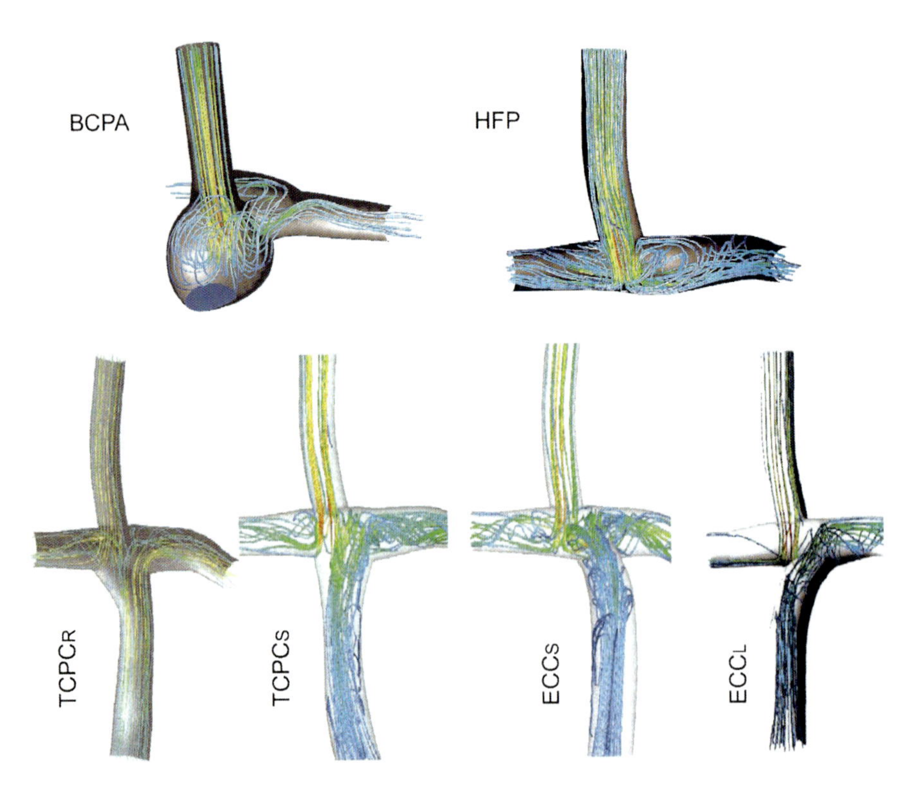

Fig. 4.19 Pathlines coloured by velocity magnitude in realistic model of bidirectional cavopulmonary anastomosis (*top left*), hemi-Fontan (*top right*), total cavopulmonary anastomosis (*bottom left*) and extracardiac conduits (*bottom right*) (The images are modified from Bove et al. (2003))

syndrome (Bove et al. 2003). In this study, the hydraulic performance of the hemi-Fontan (HFP) was compared to the bidirectional Glenn (BCPA) procedures. A three-dimensional model was constructed for typical HFP and BCPA operations based on the anatomic data derived from MRI scans, angiography and echocardiography (Fig. 4.19). Fontan models of the lateral tunnel, TCPC and extracardiac conduit (ECC) were constructed. The HFP and BCPA procedures demonstrated nearly identical performance with similar power loss and comparable flow distribution to lung. The lateral tunnel Fontan following HFP had a lower power loss compared to TCPC. The findings of this study indicated that the hemodynamic difference between the two commonly used methods of HFP and BDG is only realized after the completion Fontan operation. Similarly, Amodeo et al. (2002) concluded that the total extracardiac cavopulmonary connection with left-sided diversion of the IVC conduit anastomosis results in a central vortex that regulates the caval flow partitioning, and provides a more favorable energy-saving pattern than those with the total extracardiac cavopulmonary connection with directly opposed cavopulmonary anastomoses (Fig. 4.20).

Computer models have played an important role in attaining the optimal configuration for cavopulmonary connections. Geometric designs with minimal energy loss

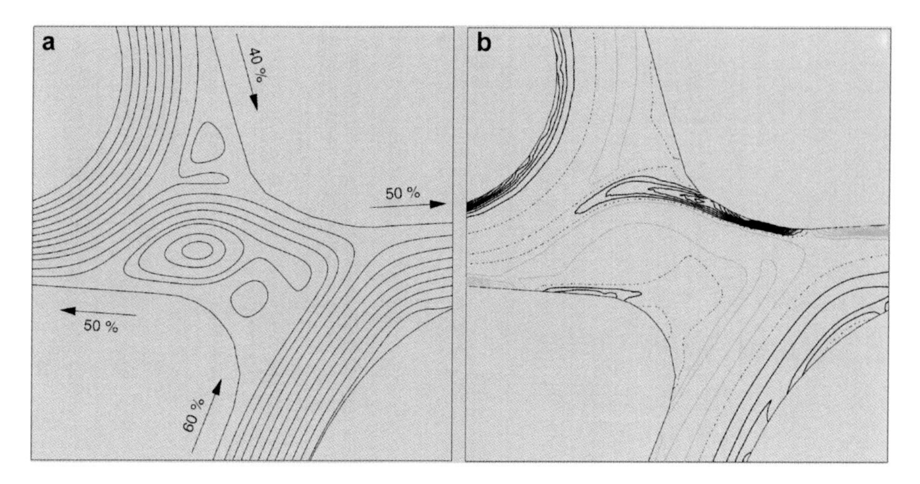

Fig. 4.20 Computed flow field in a TCPC configuration (*Re* 700, IVC 60% of total venous return, SVC 40%, equal pulmonary arterial flows). (**a**) Streamlines. (**b**) Vorticity field contours, level from zero (*dotted lines*) with steps at ±0.5. *Grey lines* represent positive levels; *black lines* represent negative levels (Modified from Amodeo et al. (2002))

are likely to lead to improved clinical outcomes. However, idealized computer models must be cautiously applied to each clinical situation, as individual anatomic variations may result in significantly different outcomes.

4.4.2 Anastomoses

An anastomosis is a surgical connection between two biological and/or synthetic conduits inside the body. The term most commonly refers to a connection, which is created between tubular structures, such as blood vessels. A pathological anastomosis may result from trauma or disease, and can involve either veins or arteries. These are usually referred to as fistulas. Generally, the anastomoses can be categorized into three classes: (1) *end-to-end*, (2) *end-to-side*, and (3) *side-to-side* (Fig. 4.21). The *end-to-end* implies that the two conduits are sutured along their transversal diameters; the *end-to-side* term indicates that the host vessel is sutured laterally to a parent vessel by means of a longitudinal incision, and the *side-to-side* specifies that the two vessels are joined along their longitudinal direction.

The local fluid dynamics at anastomoses has become a concern in the pathogenesis of atherosclerotic diseases, since the atherosclerotic plaques tend to be localized at sites of branching, and arterial curvature where the flow is disturbed. Based on this observation, several studies suggested that the local fluid dynamics may play an initiating role in intimal hyperplasia (IH) and atherosclerosis (Kabinejadian et al. 2010; Kleinstreuer et al. 2001; Haruguchi and Teraoka 2003). The performance of a surgical anastomosis depends on compliance mismatch, new geometries and consequently new hemodynamics. Indeed, increased platelet deposition at the site of

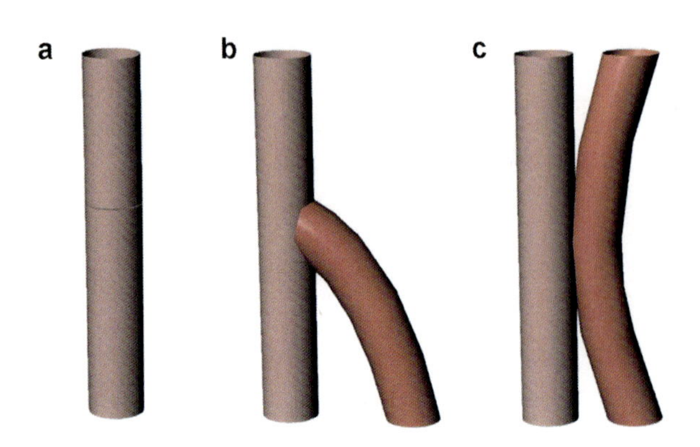

Fig. 4.21 Different types of anastomoses: (**a**) end-to-end; (**b**) end-to-side; (**c**) side-to-side (Modified from Migliavacca and Dubini 2005)

anastomosis and faster re-occlusion of the graft are phenomena that are related to the vascular fluid dynamics and the interaction with compliant vascular structure. Therefore, optimization of the anastomosis from both fluid dynamics and structural standpoints (i.e. physiological wall shear stresses, minimization of vortices and secondary flows, reduction of compliance mismatch) is a major target for computational studies.

Due to the important role of local fluid dynamics in intimal hyperplasia, a complementary approach to investigate vortex formation and evolution at anastomoses may include the numerical simulation of mass transport of blood-borne elements (such as platelets and monocytes) interacting with an intact, but potentially dysfunctional, arterial wall endothelium at the sites where IH initializes. sites of initialization of IH. Longest and Kleinstreuer (2003) performed a one-way coupled study to calculate trajectories of critical blood particles, such as platelets, on a Lagrangian basis in realistic femoro-popliteal distal anastomoses (Fig. 4.22). Platelet trajectories in Fig. 4.22d highlight the significant axial vortices in the proximal and distal regions elicited by the abrupt anastomotic profile. Significant recirculation within the junction region and highly focal separation area along the arterial floor are evident. In peripheral anastomoses, use of dedicated devices can completely transform the observed vortex pattern. Heise et al. (2004) used PIV to investigate Silastic models of a Taylor patch, a Miller cuff and a femoro-crural patch prosthesis under pulsatile conditions. Flow visualization indicates that in the Taylor patch model a jet of fluid appears to discharge into the wide anastomosis. In the Miller cuff model, a high-velocity mainstream enters the center of anastomosis dividing towards the outlets. In conclusions, significant recirculation and energy dissipation are commonly associated with surgical anastomoses. Methods to minimize their spatial and temporal extension would help better perfusion of distal organs and prevent the progression intimal hyperplasia.

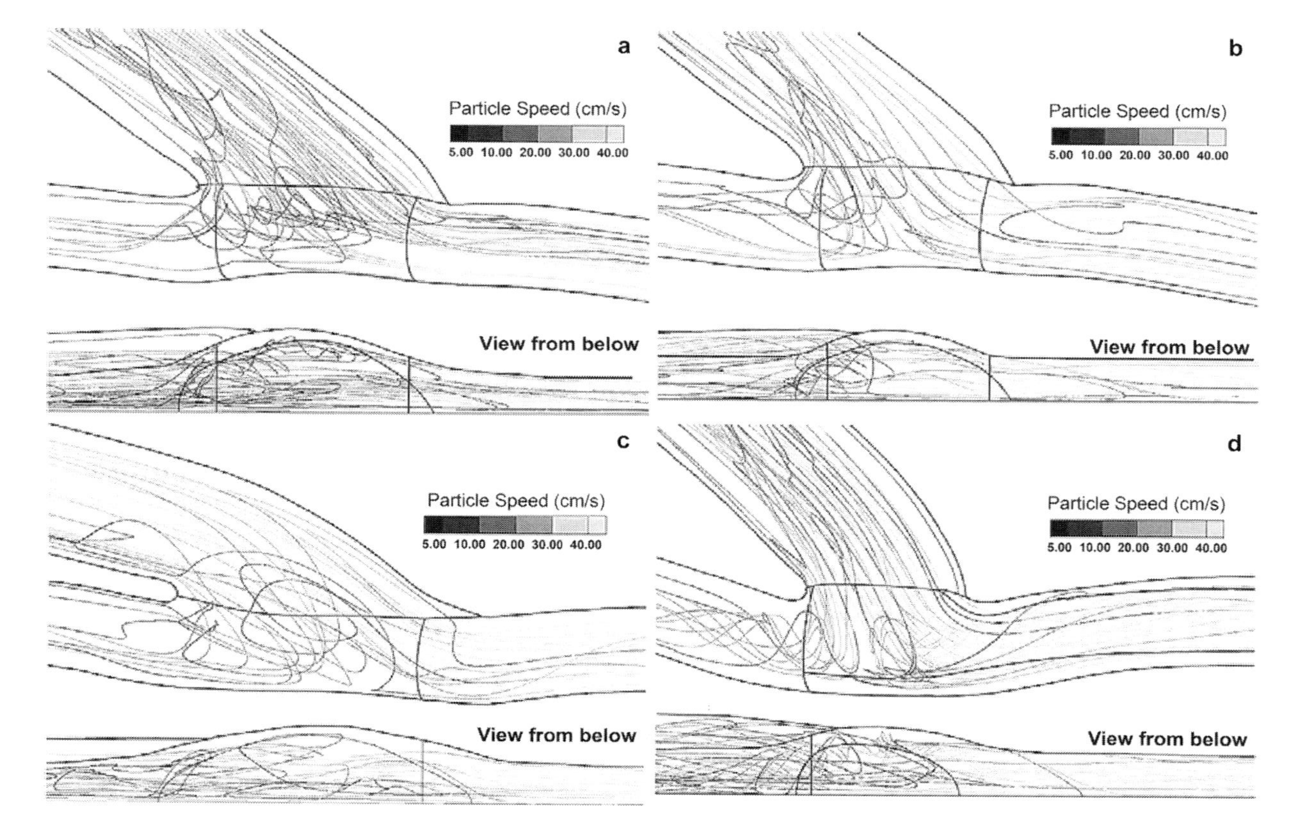

Fig. 4.22 Selected platelet trajectories indicating transient vortical flow features for four different peripheral grafts models (**a–d**) (The images are adapted from Longest and Kleinstreuer (2003))

References

Abu-Omar Y, Ratnatunga CP. Prosthetic heart valves. Surgery (Oxford). 2008;26(12):496–500.

Affeld K. Cardiac valves and cardiac assist systems. In: Fluid dynamic and biological flows, lecture series. Belgium: Von Karman Institute for Fluid Dynamics. 1998. p. 1–10.

Affeld K, Ziemann A, Goubergrits L. Technical obstacles on the road towards a permanent left ventricular assist device. Heart Vessels. 1997; (Suppl) 12:28–30.

Akutsu T, Higuchi D. Effect of the mechanical prosthetic mono-and bileaflet heart valve orientation on the flow field inside the simulated ventricle. J Artif Organs. 2000;3(2):126–35.

Allen GS, Murray KD, Olsen DB. The importance of pulsatile and nonpulsatile flow in the design of blood pumps. Artif Organs. 1997;21(8):922–8.

Aluri S, Chandran KB. Numerical simulation of mechanical mitral heart valve closure. Ann Biomed Eng. 2001;29(8):665–76.

Amodeo A et al. The beneficial vortex and best spatial arrangement in total extracardiac cavopulmonary connection. J Thorac Cardiovasc Surg. 2002;124(3):471–8.

Apte SS et al. Current developments in the tissue engineering of autologous heart valves: moving towards clinical use. Future Cardiol. 2011;7(1):77–97.

Argueta-Morales IR et al. Use of computational fluid dynamics (CFD) to tailor the surgical implantation of a ventricular assist device (VAD): a patient-specific approach to reduce risk of stroke. J Am Coll Surg. 2010;211(3, Supplement 1):S26–7.

Avrahami I. The effect of structure on the hemodynamics of artificial blood pumps. Tel Aviv: Tel Aviv University; 2003. p. 168.

Avrahami I et al. Can vortices in the flow across mechanical heart valves contribute to cavitation? Med Biol Eng Comput. 2000;38(1):93–7.

Avrahami I, et al. CFD analysis of flow through mechanical heart valves. In: IV world congress biomechanics. Calgary; 2002.

Avrahami I, Rosenfeld M, Einav S. The hemodynamics of the berlin pulsatile VAD and the role of its MHV configuration. Ann Biomed Eng. 2006a;34(9):1373–88.

Avrahami I et al. Numerical model of flow in a sac-type ventricular assist device. Artif Organs. 2006b;30(7):529–38.

Bachmann C et al. Fluid dynamics of a pediatric ventricular assist device. Artif Organs. 2000; 24(5):362–72.

Bakhtiary F et al. Impact of patient-prosthesis mismatch and aortic valve design on coronary flow reserve after aortic valve replacement. J Am Coll Cardiol. 2007;49(7):790–6.

Baldwin JT et al. LDA measurements of mean velocity and Reynolds stress fields within an artificial heart ventricle. J Biomech Eng. 1994;116(2):190–200.

Bang JS, Choi CR, Kim CN. A numerical analysis on the curved bileaflet mechanical heart valve (MHV): leaflet motion and blood flow in an elastic blood vessel. J Mech Sci Technol. 2005; 19(9):1761–72.

Behbahani M et al. A review of computational fluid dynamics analysis of blood pumps. Eur J Appl Math. 2009;20(04):363–97.

Birks EJ. Left ventricular assist devices. Heart. 2010;96(1):63–71.

Bluestein D, Li YM, Krukenkamp IB. Free emboli formation in the wake of bi-leaflet mechanical heart valves and the effects of implantation techniques. J Biomech. 2002;35(12):1533–40.

Bluestein D, Chandran KB, Manning KB. Towards non-thrombogenic performance of blood recirculating devices. Ann Biomed Eng. 2010;38(3):1236–56.

Borazjani I, Ge L, Sotiropoulos F. High-resolution fluid–structure interaction simulations of flow through a bi-leaflet mechanical heart valve in an anatomic aorta. Ann Biomed Eng. 2010;38(2): 326–44.

Bove EL et al. Computational fluid dynamics in the evaluation of hemodynamic performance of cavopulmonary connections after the norwood procedure for hypoplastic left heart syndrome. J Thorac Cardiovasc Surg. 2003;126(4):1040–7.

Brujan EA. Cardiovascular cavitation. Med Eng Phys. 2009;31(7):742–51.

Cannegieter SC, Rosendaal FR, Briet E. Thromboembolic and bleeding complications in patients with mechanical heart-valve prostheses. Circulation. 1994;89(2):635–41.
Chambers JB et al. Nominal size in six bileaflet mechanical aortic valves: a comparison of orifice size and biologic equivalence. J Thorac Cardiovasc Surg. 2003;125(6):1388–93.
Chandran KB, Rittgers SE, Yoganathan AP. Biofluid mechanics: the human circulation. Boca Raton: CRC Press; 2006.
Cheng R, Lai YG, Chandran KB. Three-dimensional fluid-structure interaction simulation of bileaflet mechanical heart valve flow dynamics. Ann Biomed Eng. 2004;32(11):1471–83.
Choi CR, Kim CN. Numerical analysis on the hemodynamics and leaflet dynamics in a bileaflet mechanical heart valve using a fluid-structure interaction method. ASAIO J. 2009;55(5): 428–37.
Christensen CW et al. Evaluating the flow characteristics of artificial pumping devices using nuclear scintigraphy. Artif Organs. 1993;17(10):843–9.
Clegg AJ et al. Clinical and cost-effectiveness of left ventricular assist devices as a bridge to heart transplantation for people with end-stage heart failure: a systematic review and economic evaluation. Eur Heart J. 2006;27(24):2929–38.
Cooper BT et al. The influence of operational protocol on the fluid dynamics in the 12 cc Penn state pulsatile pediatric ventricular assist device: the effect of end-diastolic delay. Artif Organs. 2010;34(4):E122–33.
Copeland JG et al. Cardiac replacement with a total artificial heart as a bridge to transplantation [see comment]. N Engl J Med. 2004;351(9):859–67.
d'Arcy JL et al. Valvular heart disease: the next cardiac epidemic. Heart. 2011;97(2):91–3.
da Vinci L. Quademi d'Anatomica II. 1513:9.
Daneshmand MA et al. Left ventricular assist device destination therapy versus extended criteria cardiac transplant. Ann Thorac Surg. 2010;89(4):1205–10.
Dasi LP et al. Fluid mechanics of artificial heart valves. Clin Exp Pharmacol Physiol. 2009; 36(2):225–37.
de Leval MR et al. Modified Blalock-Taussig shunt. Use of subclavian artery orifice as flow regulator in prosthetic systemic-pulmonary artery shunts. J Thorac Cardiovasc Surg. 1981;81(1):112–9.
de Leval MR et al. Total cavopulmonary connection: a logical alternative to atriopulmonary connection for complex Fontan operations. Experimental studies and early clinical experience. J Thorac Cardiovasc Surg. 1988;96(5):682–95.
de Leval MR et al. Use of computational fluid dynamics in the design of surgical procedures: application to the study of competitive flows in cavopulmonary connections. J Thorac Cardiovasc Surg. 1996;111(3):502–13.
de Tullio MD, Pedrizzetti G, Verzicco R. On the influence of the aortic root geometry on blood flow after a bileaflet mechanical heart valve implant: a numerical study. Acta Mechanica 2011; 216:147–163.
de Zelicourt DA et al. In vitro flow analysis of a patient-specific intraatrial total cavopulmonary connection. Ann Thorac Surg. 2005;79(6):2094–102.
Deklunder G et al. Can cerebrovascular microemboli induce cognitive impairment in patients with prosthetic heart valves? Eur J Ultrasound. 1998;7(1):47–51.
Deutsch S et al. Experimental fluid mechanics of pulsatile artificial blood pumps. Annu Rev Fluid Mech. 2006;38:65–86.
Douglas WI et al. Hemi-Fontan procedure for hypoplastic left heart syndrome: outcome and suitability for Fontan. Ann Thorac Surg. 1999;68(4):1361–7.
Doyle MG et al. Numerical simulations of blood flow in artificial and natural hearts with fluid-structure interaction. Artif Organs. 2008;32(11):870–9.
Dubini G et al. A numerical fluid mechanical study of repaired congenital heart defects. Application to the total cavopulmonary connection. J Biomech. 1996;29(1):111–21.
Dumont K et al. Comparison of the hemodynamic and thrombogenic performance of two bileaflet mechanical heart valves using a CFD/FSI model. J Biomech Eng. 2007;129:558.
Eichler MJ, Reul HM. Mechanical heart valve cavitation: valve specific parameters. Int J Artif Organs. 2004;27(10):855–67.

Einav S, et al. Numerical and experimental measurements of the flow through mechanical heart valves in the natural and artificial heart. In: Second joint EMBS-BMES conference 2002. 24th annual international conference of the Engineering in Medicine and Biology Society. Annual fall meeting of the Biomedical Engineering Society (Cat. No.02CH37392). Houston; 2002.

Ellis JT, Travis BR, Yoganathan AP. An in vitro study of the hinge and near-field forward flow dynamics of the St. Jude Medical (R) Regent (TM) bileaflet mechanical heart valve. Ann Biomed Eng. 2000;28(5):524–32.

Escobedo C, Tovar F, Suárez B, Hernández A, Corona F, Sacristán E. Experimental and computer based performance analysis of two elastomer VAD valve designs. IEEE EMBS 2005;7620:3. doi: 10.1109/IEMBS.2005.1616276.

Faludi R et al. Left ventricular flow patterns in healthy subjects and patients with prosthetic mitral valves: an in vivo study using echocardiographic particle image velocimetry. J Thorac Cardiovasc Surg. 2010;139(6):1501–10.

Fan H-m et al. Applications of CFD technique in the design and flow analysis of implantable axial flow blood pump. J Hydrodyn Ser B. 2010;22(4):518–25.

Fang JC. Rise of the machines–left ventricular assist devices as permanent therapy for advanced heart failure. N Engl J Med. 2009;361(23):2282.

Garon A, Farinas MI. Fast three-dimensional numerical hemolysis approximation. Artif Organs. 2004;28(11):1016–25.

Ge L et al. Flow in a mechanical bileaflet heart valve at laminar and near-peak systole flow rates: CFD simulations and experiments. J Biomech Eng. 2005;127(5):782–97.

Gharib M et al. Leonardo's vision of flow visualization. Exp Fluids. 2002;33(1):219–23.

Goetze S et al. In vivo short-term doppler hemodynamic profiles of 189 carpentier-edwards perimount pericardial bioprosthetic valves in the mitral position. J Am Soc Echocardiogr. 2004;17(9):981–7.

Goldstein DJ, Oz MC, Rose EA. Implantable left ventricular assist devices. N Engl J Med. 1998; 339(21):1522–33.

Goubergrits L et al. Numerical dye washout method as a tool for characterizing the heart valve flow: Study of three standard mechanical heart valves. ASAIO J. 2008;54(1):50–7.

Govindarajan V, Udaykumar HS, Chandran KB. Flow dynamic comparison between recessed hinge and open pivot bi-leaflet heart valve designs. J Mech Med Biol. 2009a;9(2):161–76.

Govindarajan V, Udaykumar HS, Chandran KB. Two-dimensional simulation of flow and platelet dynamics in the hinge region of a mechanical heart valve. J Biomech Eng. 2009b;131(3): 031002.

Grigioni M et al. The influence of the leaflets' curvature on the flow field in two bileaflet prosthetic heart valves. J Biomech. 2001;34(5):613–21.

Gross JM et al. Microstructural flow analysis within a bileaflet mechanical heart valve hinge. J Heart Valve Dis. 1996;5(6):581–90.

Haller JAJ, Adkins JC, Worthington M, Rauenhorst J. Experimental studies on permanent bypass of the right heart. Surgery. 1966;59(6):1128–32.

Haruguchi H, Teraoka S. Intimal hyperplasia and hemodynamic factors in arterial bypass and arteriovenous grafts: a review. J Artif Organs. 2003;6(4):227–35.

Heise M et al. Flow pattern and shear stress distribution of distal end-to-side anastomoses. A comparison of the instantaneous velocity fields obtained by particle image velocimetry. J Biomech. 2004;37(7):1043–51.

Herbertson LH, Deutsch S, Manning KB. Modifying a tilting disk mechanical heart valve design to improve closing dynamics. J Biomech Eng. 2008;130(5):054503.

Hochareon P et al. Correlation of in vivo clot deposition with the flow characteristics in the 50 cc Penn state artificial heart: a preliminary study. ASAIO J. 2004a;50(6):537–42.

Hochareon P et al. Wall shear-rate estimation within the 50 cc Penn state artificial heart using particle image velocimetry. J Biomech Eng. 2004b;126(4):430–7.

Hoffmann G, Lutter G, Cremer J. Durability of bioprosthetic cardiac valves. Dtsch Arztebl Int. 2008;105(8):143–8.

Holman WL et al. Device related infections: are we making progress? J Card Surg. 2010; 25(4):478–83.
Hose DR et al. Fundamental mechanics of aortic heart valve closure. J Biomech. 2006; 39(5):958–67.
Iwasaki K et al. Development of a polymer bileaflet valve to realize a low-cost pulsatile blood pump. Artif Organs. 2003;27(1):78–83.
Johansen P. Mechanical heart valve cavitation. Expert Rev Med Devices. 2004;1(1):95–104.
John R et al. Physiologic and pathologic changes in patients with continuous-flow ventricular assist devices. J Cardiovasc Transl Res. 2009;2(2):154–8.
Kabinejadian F et al. A novel coronary artery bypass graft design of sequential anastomoses. Ann Biomed Eng. 2010;38(10):3135–50.
Kaminsky R et al. Flow visualization through two types of aortic prosthetic heart valves using stereoscopic high-speed particle image velocimetry. Artif Organs. 2007;31:869–79.
Katayama S et al. The sinus of valsalva relieves abnormal stress on aortic valve leaflets by facilitating smooth closure. J Thorac Cardiovasc Surg. 2008;136(6):1528–35.
Kawahito K, Adachi H, Ino T. Platelet activation in the gyro C1E3 centrifugal pump: comparison with the terumo capiox and the Nikkiso HPM-15. Artif Organs. 2000;24(11):889–92.
Kelly SG. Computational fluid dynamics insights in the design of mechanical heart valves. Artif Organs. 2002;26(7):608–13.
Kheradvar A, Gharib M. On mitral valve dynamics and its connection to early diastolic flow. Ann Biomed Eng. 2009;37(1):1–13.
Kheradvar A, Kasalko J, Johnson D, Gharib M. An in vitro study of changing profile heights in mitral bioprostheses and their influence on flow. ASAIO J. 2006;52(1):34–8.
Kheradvar A, Houle H, Pedrizzetti G, Tonti G, Belcik T, Ashraf M, Lindner JR, Gharib M, Sahn DJ. Echocardiographic particle image velocimetry: a novel technique for quantification of left ventricular blood vorticity pattern. J Am Soc Echocardiogr. 2010;23(1):86–94.
Kheradvar A, Falahatpisheh, A. The Effects of Dynamic Saddle Annulus and Leaflet Length on Transmitral Flow Pattern and Leaflet Stress of a Bi-leaflet Bioprosthetic Mitral Valve, Journal of Heart Valve Disease, 2012 in press.
Khunatorn Y, Mahalingam S, DeGroff CG, Shandas R. Influence of connection geometry and SVC-IVC flow rate ratio on flow structures within the total cavopulmonary connection: a numerical study. J Biomech Eng. 2002;124(4):364–77.
Kini V et al. Integrating particle image velocimetry and laser Doppler velocimetry measurements of the regurgitant flow field past mechanical heart valves. Artif Organs. 2001;25(2):136–45.
Kleine P et al. Effect of mechanical aortic valve orientaion on coronary artery flow: comparison of tilting disc versus bileaflet prostheses in pigs. J Thorac Cardiovasc Surg. 2002;124(5):925–32.
Kleinstreuer C, Hyun S, Buchanan Jr JR, Longest PW, Archie Jr JP, Truskey GA. Hemodynamic parameters and early intimal thickening in branching blood vessels. Crit Rev Biomed Eng. 2001;29(1):1–64.
Knapp Y, Bertrand E. Particle imaging velocimetry measurements in a heart simulator. J Vis. 2005; 8(3):217–24.
Koenig CS, Clark C. Flow mixing and fluid residence times in a model of a ventricular assist device. Med Eng Phys. 2001;23(2):99–110.
Korakianitis T, Shi Y. Numerical comparison of hemodynamics with atrium to aorta and ventricular apex to aorta VAD support. ASAIO J. 2007;53(5):537.
Kreider JW et al. The 50 cc Penn state left ventricular assist device: a parametric study of valve orientation flow dynamics. ASAIO J. 2006;52(2):123–31.
Krishnan S et al. Two-dimensional dynamic simulation of platelet activation during mechanical heart valve closure. Ann Biomed Eng. 2006;34(10):1519–34.
Kvitting JPE et al. In vitro assessment of flow patterns and turbulence intensity in prosthetic heart valves using generalized phase-contrast MRI. J Magn Reson Imaging. 2010;31(5):1075–80.
Laas J et al. Orientation of tilting disc and bileaflet aortic valve substitutes for optimal hemodynamics. Ann Thorac Surg. 1999;68(3):1096–9.

Lahpor JR. State of the art: implantable ventricular assist devices. Curr Opin Organ Transplant. 2009;14(5):554–9.
Landesberg A et al. Effects of synchronized cardiac assist device on cardiac energetics. Ann N Y Acad Sci. 2006;1080:466–78. Interactive and Integrative Cardiology.
Lee H, Homma A, Taenaka Y. Hydrodynamic characteristics of bileaflet mechanical heart valves in an artificial heart: cavitation and closing velocity. Artif Organs. 2007;31(7):532–7.
Lee H et al. Effects of leaflet geometry on the flow field in three bileaflet valves when installed in a pneumatic ventricular assist device. J Artif Organs. 2009;12(2):98–104.
Lee H et al. Observation of cavitation pits on mechanical heart valve surfaces in an artificial heart used in in vitro testing. J Artif Organs. 2010;13(1):17–23.
Leo H-L et al. A comparison of flow field structures of two tri-leaflet polymeric heart valves. Ann Biomed Eng. 2005;33(4):429–43.
Li CP et al. Role of vortices in cavitation formation in the flow across a mechanical heart valve. J Heart Valve Dis. 2008;17(4):435–45.
Li CP, Lo CW, Lu PC. Estimation of viscous dissipative stresses induced by a mechanical heart valve using PIV data. Ann Biomed Eng. 2010;38(3):903–16.
Lietz K et al. Outcomes of left ventricular assist device implantation as destination therapy in the post-REMATCH era: implications for patient selection. Circulation. 2007;116(5):497–505.
Lim WL et al. Cavitation phenomena in mechanical heart valves: the role of squeeze flow velocity and contact area on cavitation initiation between two impinging rods. J Biomech. 2003;36(9):1269–80.
Longest PW, Kleinstreuer C. Particle-hemodynamics modeling of the distal end-to-side femoral bypass: effects of graft caliber and graft-end cut. Med Eng Phys. 2003;25(10):843–58.
Machler H et al. Influence of bileaflet prosthetic mitral valve orientation on left ventricular flow-an experimental in vivo magnetic resonance imaging study. Eur J Cardiothorac Surg. 2004; 26(4):747–53.
Machler H et al. Influence of a tilting prosthetic mitral valve orientation on the left ventricular flow-an experimental in vivo magnetic resonance imaging study. Eur J Cardiothorac Surg. 2007;32(1):102–7.
Mahmood AK et al. Critical review of current left ventricular assist devices. Perfusion. 2000; 15(5):399–420.
Maines BH, Brennen CE. Lumped parameter model for computing the minimum pressure during mechanical heart valve closure. J Biomech Eng. 2005;127(4):648–55.
Maire R et al. Abnormalities of left-ventricular flow following mitral-valve replacement-a color-flow Doppler study. Eur Heart J. 1994;15(3):293–302.
Manning KB et al. A detailed fluid mechanics study of tilting disk mechanical heart valve closure and the implications to blood damage. J Biomech Eng. 2008;130(4):041001-1–8.
Markl M et al. Three-dimensional magnetic resonance flow analysis in a ventricular assist device. J Thorac Cardiovasc Surg. 2007;134(6):1471–6.
Martin J et al. Improved durability of the HeartMate XVE left ventricular assist device provides safe mechanical support up to 1 year but is associated with high risk of device failure in the second year. J Heart Lung Transplant. 2006;25(4):384–90.
Maymir JC et al. Effects of tilting disk heart valve gap width on regurgitant flow through an artificial heart mitral valve. Artif Organs. 1997;21(9):1014–25.
May-Newman K, Hillen B, Dembitsky W. Effect of left ventricular assist device outflow conduit anastomosis location on flow patterns in the native aorta. ASAIO J. 2006;52(2):132–9.
May-Newman K et al. Biomechanics of the aortic valve in the continuous flow VAD-assisted heart. ASAIO J. 2010;56(4):301–8. doi:10.1097/MAT.0b013e3181e321da.
Medart D et al. PIV investigation in the hinge area of a novel bileaflet mechanical heart valve (HIA-bileaflet). Int J Artif Organs. 2004;27(7):605.
Medvitz RB et al. Validation of a CFD methodology for positive displacement LVAD analysis using PIV data. J Biomech Eng. 2009;131(11):111009.
Meuris B, Verbeken E, Flameng W. Mechanical valve thrombosis in a chronic animal model: differences between monoleaflet and bileaflet valves. J Heart Valve Dis. 2005;14(1):96–104.

Meyer RS et al. Laser Doppler velocimetry and flow visualization studies in the regurgitant leakage flow region of three mechanical mitral valves. Artif Organs. 2001;25(4):292–9.

Migliavacca F, de Leval MR, Dubini G, Pietrabissa R. A computational pulsatile model of the bidirectional cavopulmonary anastomosis: the influence of pulmonary forward flow. J Biomech Eng. 1996;118(4):520–8.

Migliavacca F et al. Computational transient simulations with varying degree and shape of pulmonic stenosis in models of the bidirectional cavopulmonary anastomosis. Med Eng Phys. 1997;19(4):394–403.

Migliavacca F, Kilner PJ, Pennati G, Dubini G, Pietrabissa R, Fumero R, et al. Computational fluid dynamic and magnetic resonance analyses of flow distribution between lungs after total cavopulmonary connection. IEEE Trans Biomed Eng. 1999;46(4):393–9.

Migliavacca F, Dubini G. 2005. Computational modeling of vascular anastomoses. Biomechan Model Mechanobiol 3:235–250

Miller LW. The development of the von Willebrand syndrome with the use of continuous flow left ventricular assist devices: a cause-and-effect relationship. J Am Coll Cardiol. 2010;56: 1214–1215.

Milo S et al. Mitral mechanical heart valves: in vitro studies of their closure, vortex and microbubble formation with possible medical implications. Eur J Cardiothorac Surg. 2003;24(3): 364–70.

Minami K et al. Morbidity and outcome after mechanical ventricular support using Thoratec, Novacor, and HeartMate for bridging to heart transplantation. Artif Organs. 2000;24(6):421–6.

Mohammadi H, Ahmadian MT, Wan WK. Time-dependent analysis of leaflets in mechanical aortic bileaflet heart valves in closing phase using the finite strip method. Med Eng Phys. 2006;28(2):122–33.

Moosavi MH, Fatouraee N, Katoozian H. Finite element analysis of blood flow characteristics in a Ventricular Assist Device (VAD). Simul Model Pract Theory. 2009;17(4):654–63.

Morbiducci U et al. Blood damage safety of prosthetic heart valves. Shear-induced platelet activation and local flow dynamics: a fluid-structure interaction approach. J Biomech. 2009;42(12): 1952–60.

Mouret F et al. Mitral prosthesis opening and flow dynamics in a model of left ventricle: an in vitro study on a monoleaflet mechanical valve. Cardiovasc Eng. 2005;5(1):13–20.

Mussivand T, Day KD, Naber BC. Fluid dynamic optimization of a ventricular assist device using particle image velocimetry. ASAIO J. 1999;45(1):25–31.

Naftali S, Avrahami I,Landesberg A. Quantification of the hemodynamics inside a novel synchronized therapeutic cardiac assist device for chronic heart failure. IEEE ITRE 2006; 116–20. doi: 10.1109/ITRE.2006.381546.

Nkomo VT et al. Burden of valvular heart diseases: a population-based study. Lancet. 2006;368(9540):1005–11.

Nobili M et al. Numerical simulation of the dynamics of a bileaflet prosthetic heart valve using a fluid-structure interaction approach. J Biomech. 2008;41(11):2539–50.

Nose Y et al. Development of rotary blood pump technology: past, present, and future. Artif Organs. 2000;24(6):412–20.

Ohta Y et al. Effect of the sinus of valsalva on the closing motion of bileaflet prosthetic heart valves. Artif Organs. 2000;24(4):309–12.

Okamoto E et al. Blood compatible design of a pulsatile blood pump using computational fluid dynamics and computer-aided design and manufacturing technology. Artif Organs. 2003; 27(1):61–7.

Okamoto E et al. Development of integrated electronics unit for drive and control of undulation pump-left ventricular assist device. Artif Organs. 2006;30(5):403–5.

Oley LA et al. Off-design considerations of the 50 cc Penn state ventricular assist device. Artif Organs. 2005;29(5):378–86.

Osaki S et al. Improved survival in patients with ventricular assist device therapy: the University of Wisconsin experience. Eur J Cardiothorac Surg. 2008;34(2):281.

Pagani FD et al. Extended mechanical circulatory support with a continuous-flow rotary left ventricular assist device. J Am Coll Cardiol. 2009;54(4):312–21.
Pai CN, Shinshi T, Shimokohbe A. Sensorless measurement of pulsatile flow rate using a disturbance force observer in a magnetically levitated centrifugal blood pump during ventricular assistance. Flow Meas Instrum. 2010;21(1):33–9.
Park SJ et al. Left ventricular assist devices as destination therapy: a new look at survival. J Thorac Cardiovasc Surg. 2005;129(1):9–17.
Peacock JA. An in vitro study of the onset of turbulence in the sinus of valsalva. Circ Res. 1990; 67(2):448–60.
Pedrizzetti G, Domenichini F, Tonti G. On the left ventricular vortex reversal after mitral valve replacement. Ann Biomed Eng. 2010;38(3):769–73.
Pop G et al. What is the ideal orientation of a mitral disk prosthesis-an invivo hemodynamic-study based on color flow imaging and continuous wave Doppler. Eur Heart J. 1989;10(4):346–53.
Querzoli G, Fortini S, Cenedese A. Effect of the prosthetic mitral valve on vortex dynamics and turbulence of the left ventricular flow. Phys Fluids. 2010;22(4):041901–10.
Rahimtoola SH. Choice of prosthetic heart valve in adults an update. J Am Coll Cardiol. 2010;55(22):2413–26.
Rambod E et al. Role of vortices in growth of microbubbles at mitral mechanical heart valve closure. Ann Biomed Eng. 2007;35(7):1131–45.
Raz S, et al. Experimental analysis of flow through tilting disk valve. In: Conference proceedings. IV world congress biomechanics. Calgary; 2002.
Raz S et al. DPIV prediction of flow induced platelet activation-comparison to numerical predictions. Ann Biomed Eng. 2007;35(4):493–504.
Rose ML, Mackay TG, Wheatley DJ. Evaluation of four blood pump geometries: fluorescent particle flow visualisation technique. Med Eng Phys. 2000;22(3):201–14.
Rose EA et al. Randomized Evaluation of Mechanical Assistance for the Treatment of Congestive Heart Failure (REMATCH) Study Group, "Long-term mechanical left ventricular assistance for end-stage heart failure. N Engl J Med. 2001;345(20):1435–43.
Rosenfeld M, Avrahami I, Einav S. The time-dependent flow across a model of a mitral tilting disk valve and the left ventricle. In: ASME 1999 bioengineering conference. Big Sky; 1999.
Rosenfeld M, Avrahami I, Einav S. Unsteady effects on the flow across tilting disk valves. J Biomech Eng. 2002;124(1):21–9.
Rothenburger M et al. Treatment of thrombus formation associated with the MicroMed DeBakey VAD using recombinant tissue plasminogen activator. Circulation. 2002;106(90121):I-189–92.
Ryu K et al. Importance of accurate geometry in the study of the total cavopulmonary connection: computational simulations and *in vitro* experiments. Ann Biomed Eng. 2001;29(10):844–53.
Saito S et al. Risk factor analysis of long-term support with left ventricular assist system. Circ J. 2010;74(4):715–22.
Sakota D et al. Mechanical damage of red blood cells by rotary blood pumps: selective destruction of aged red blood cells and subhemolytic trauma. Artif Organs. 2008;32(10):785–91.
Sano S et al. Right ventricle-pulmonary artery shunt in first-stage palliation of hypoplastic left heart syndrome. J Thorac Cardiovasc Surg. 2003;126(2):504–10.
Sato K et al. Analysis of flow patterns in a ventricular assist device: a comparative study of particle image velocimetry and computational fluid dynamics. Artif Organs. 2009;33(4):352–9.
Saxena R et al. An in vitro assessment by means of laser Doppler velocimetry of the medtronic advantage bileaflet mechanical heart valve hinge flow. J Thorac Cardiovasc Surg. 2003;126(1):90–8.
Schoephoerster RT, Chandran KB. Velocity and turbulence measurements past mitral-valve prostheses in a model left-ventricle. J Biomech. 1991;24(7):549–62.
Senthilnathan V, Treasure T, Grunkemeier G, Starr A. Heart valves: which is the best choice? Cardiovasc Surg. 1999;7:393–7.
Sezai A et al. Major organ function under mechanical support: comparative studies of pulsatile and nonpulsatile circulation. Artif Organs. 1999;23(3):280–5.

Shapira Y, Vaturi M, Sagie A. Hemolysis associated with prosthetic heart valves a review. Cardiol Rev. 2009;17(3):121–4.

Shu MCS et al. Flow characterization of the ADVANTAGE (R) and St. Jude Medical (R) bileaflet mechanical heart valves. J Heart Valve Dis. 2004;13(5):814–22.

Simon HA et al. Comparison of the hinge flow fields of two bileaflet mechanical heart valves under aortic and mitral conditions. Ann Biomed Eng. 2004;32(12):1607–17.

Simon HA et al. Simulation of the three-dimensional hinge flow fields of a bileaflet mechanical heart valve under aortic conditions. Ann Biomed Eng. 2010;38(3):841–53.

Skjelland M et al. Solid cerebral microemboli and cerebrovascular symptoms in patients with prosthetic heart valves. Stroke. 2008;39(4):1159–64.

Slater JP et al. Low thromboembolic risk without anticoagulation using advanced-design left ventricular assist devices. Ann Thorac Surg. 1996;62(5):1321–7. discussion 1328.

Slaughter MS et al. Advanced heart failure treated with continuous-flow left ventricular assist device. N Engl J Med. 2009;361(23):2241.

Smadi O et al. Numerical and experimental investigations of pulsatile blood flow pattern through a dysfunctional mechanical heart valve. J Biomech. 2010;43(8):1565–72.

Song G, Chua LP, Lim TM. Numerical study of a bio-centrifugal blood pump with straight impeller blade profiles. Artif Organs. 2010;34(2):98–104.

Sotiropoulos F, Borazjani I. A review of state-of-the-art numerical methods for simulating flow through mechanical heart valves. Med Biol Eng Comput. 2009;47(3):245–56.

Stephens EH et al. Age-related changes in material behavior of porcine mitral and aortic valves and correlation to matrix composition. Tissue Eng Part A. 2010;16(3):867–78.

Telman G et al. The nature of microemboli in patients with artificial heart valves. J Neuroimaging. 2002;12(1):15–8.

Timms D et al. Atrial versus ventricular cannulation for a rotary ventricular assist device. Artif Organs. 2010;34(9):714–20.

Travis B et al. Bileaflet aortic valve prosthesis pivot geometry influences platelet secretion and anionic phospholipid exposure. Ann Biomed Eng. 2001;29(8):657–64.

Travis BR et al. An analysis of turbulent shear stresses in leakage flow through a bileaflet mechanical prostheses. J Biomech Eng. 2002;124(2):155–65.

Travis AR et al. Vascular pulsatility in patients with a pulsatile-or continuous-flow ventricular assist device. J Thorac Cardiovasc Surg. 2007;133(2):517–24.

Undar A. Myths and truths of pulsatile and nonpulsatile perfusion during acute and chronic cardiac support. Artif Organs. 2004;28(5):439–43.

Uriel N et al. Acquired von Willebrand syndrome after continuous-flow mechanical device support contributes to a high prevalence of bleeding during long-term support and at the time of transplantation. J Am Coll Cardiol. 2010;56(15):1207–13.

Van Haesdonck J-M et al. Comparison by computerized numeric modeling of energy losses in different Fontan connections. Circulation. 1995;92(9):322–6.

Van Rijk-Zwikker GL, Delemarre BJ, Huysmans HA. The orientation of the bi-leaflet CarboMedics valve in the mitral position determines left ventricular spatial flow patterns. Eur J Cardiothorac Surg. 1996;10(7):513–20.

Van Steenhoven AA, Van Dongen MEH. Model studies of the closing behaviour of the aortic valve. J Fluid Mech. 1979;90(01):21–32.

Wang JH et al. Computational fluid dynamics study of a protruded-hinge bileaflet mechanical heart valve. J Heart Valve Dis. 2001;10(2):254–62.

Wu YX et al. Mechanical heart valves: are two leaflets better than one? J Thorac Cardiovasc Surg. 2004;127(4):1171–9.

Yang XL, Liu Y, Yang JM. Unsteady flow and diaphragm motion in total artificial heart. J Mech Sci Technol. 2007;21(11):1869–75.

Yin W et al. Flow-induced platelet activation in bileaflet and monoleaflet mechanical heart valves. Ann Biomed Eng. 2004;32(8):1058–66.

Yoganathan AP, He Z, Casey Jones S. Fluid mechanics of heart valves. Annu Rev Biomed Eng. 2004;6:331–62.

Yoganathan AP, Chandran KB, Sotiropoulos F. Flow in prosthetic heart valves: state-of-the-art and future directions. Ann Biomed Eng. 2005;33(12):1689–94.
Zilla P et al. Prosthetic heart valves: catering for the few. Biomaterials. 2008;29(4):385–406.
Zimpfer D et al. Long-term neurocognitive function after mechanical aortic valve replacement. Ann Thorac Surg. 2006;81(1):29–33.

Chapter 5
Diagnostic Vortex Imaging

Abstract In the field of cardiology, the current ability to image vortices is based on special applications to magnetic resonance imaging and echocardiography. In this section, we first review *in vivo* methods to visualize vortices in the cardiac chambers. Then we examine the experimental and computational techniques to model the vortex formation inside the heart, the validation schemes and characterization of vortices in the heart or cardiovascular devices.

5.1 Magnetic Resonance Imaging

5.1.1 Velocity Measurements Using MRI

5.1.1.1 Introduction to MRI

In MRI, imaging is based on the magnetization of hydrogen nuclei in the body due to a strong magnetic field (Lauterbur 1973; Haacke et al. 1999). A radio-frequency signal at the resonance frequency (Larmor frequency) excites the nuclei. When the radio-frequency signal is switched off, the nuclei relax, thereby sending a radio-wave signal, called the spin echo, which is measured by receiver coils. The Larmor frequency depends on the strength of the magnetic field. Spatial encoding can be obtained by introducing gradients on the external magnetic field during excitation and relaxation. A gradient during excitation can be used to confine the excitation to a slice or a slab. Once a gradient is present during relaxation, the spin echo will consist of different frequencies, which correspond to different locations. This is repeated with different phase-encoding gradients, in order to measure a slice or three-dimensional volume.

Real-time imaging of a beating heart is possible in cardiac MRI. Best image quality is often obtained by the use the electrocardiography (ECG) signal to gate the MR measurement from several beats, which, after Fourier transformation, is

A. Kheradvar, G. Pedrizzetti, *Vortex Formation in the Cardiovascular System*,
DOI 10.1007/978-1-4471-2288-3_5,

reconstructed into an image sequence covering a single heart beat. Two-dimensional time-resolved images are often acquired within one breath hold. Image contrast is obtained by differences in relaxation among spinning hydrogen nuclei, an effect intrinsic to the tissue type, and factors such as the type of pulse sequence used, echo time (TE) and repetition time (TR), which are defined by the operator.

Advantages of MRI include excellent anatomical detail, which can be enhanced with the aid of paramagnetic contrast agents such as gadolinium, that shorten T1 relaxation. In contrast to ultrasound, bone and lung tissue do not shadow adjacent structures and produce excellent images. Several different MRI pulse sequences exist, allowing for optimized and accurate assessment of for cardiac anatomy, motion, viability, perfusion, and blood flow (Haacke et al. 1999).

5.1.1.2 Assessment of Blood Flow with MRI

Phase contrast MRI is the gold standard for flow assessment (Moran 1982; Pelc et al. 1991; Haacke et al. 1999). In this technique, phase of the MR signal is used instead of the magnitude. Protons moving in a magnetic field gradient acquire a phase difference proportional to the velocity. Bipolar gradient lobes applied in opposite directions only result in a net phase shift for moving protons (e.g., in the blood stream). In practice, other gradients used in the acquisition scheme will also lead to a phase difference in stationary tissue. This is computed by subtracting the phase image from two measurements, one with bipolar gradients applied and one without. In this way, the velocity can be measured in any direction.

As the phase of the MR signal is used for velocity encoding, a positive phase shift larger than +180° cannot be separated from a negative phase shift in the opposite direction, larger than −180°. The range of velocities measured within the available 360° is described as the velocity encoding range (VENC), which is defined by the operator. If the velocity encoding range is set too low, high positive velocities will be falsely interpreted as high negative velocities, giving rise to the phenomenon known as *aliasing*. On the contrary, when the velocity range is set too high, the sensitivity to low flow velocities is decreased.

In most cases, one complete phase-contrast data set is reconstructed from several (15–1,000) heart cycles. Therefore, it is normal practice that heart beats with an extremely long or short R-R-interval will be disregarded to sustain sufficient data quality. Nevertheless, in the presence of severe arrhythmia, data quality will suffer.

Various disturbances to the magnetic field may cause an offset in measured velocity. Concomitant gradient effects are often corrected by the manufacturer, as they can be calculated from a prior knowledge of the gradient sequence (Bernstein et al. 1998). Offsets due to eddy currents are normally balanced in the center of the magnet by adjusting the gradient system. However, offsets due to eddy currents increase when the distance to the center increases. This can be corrected by using a least square fit to velocity in stationary tissues such as muscle and fat (Pelc et al. 1991).

It should be reminded that metallic prostheses used in orthopedic, cardiac and vascular surgery have paramagnetic properties that cause signal loss in MRI. Coronary stents, mechanical valve prostheses and sternal wires cause a signal void that often extends beyond the physical limit of the object itself.

5.1.1.3 From One-Dimensional to Three-Dimensional Flow

MRI is a diverse technique allowing for measurement of velocity along any direction, anywhere within the human body. However, the time required for measurements depends on the size and accuracy of the data measured (i.e. spatial coverage, spatial and temporal resolution), which necessitates a technique focused on type of information required, and patient's cooperation. For example, in clinical practice, the physician may only be interested in measuring the peak velocity in a blood vessel, or volume of flow across a valve. In clinical practice, a through-plane velocity-encoded time-resolved phase contrast image is often times acquired in a breath hold (Lee 2005).

The blood flow in the heart is truly three-dimensional, and can only be fully understood by a time-resolved three-directional technique. Due to the large size of data that needs to be collected, data acquisition relies on synchronization with heart rate as well as respiratory motion (Markl et al. 2011; Ebbers 2011). Respiratory motion artifacts are normally minimized by the use of respiratory gating that keeps track of the position of the diaphragm using bellows, MR navigators, or self-gating techniques. Three acquisitions sensitized to velocity, and one reference scan without bipolar gradients are required in one heart beat to obtain the three-directional blood flow velocity. Currently, the highest achievable temporal resolution is about 20 ms in humans. Temporal resolution is often decreased to reduce the measurement time. The total measurement time heavily depends on the desired data quality and resolution. Nowadays, time-resolved three-dimensional velocity data covering a complete heart with a spatial resolution of $3.0 \times 3.0 \times 3.0\ mm^3$ and temporal resolution of 46 ms can be obtained within about 20 min (Markl et al. 2011). Accelerating the acquisition by reduced sampling is possible but may induce artifacts that complicate data processing.

5.1.2 Visualization and Quantification

Clinically, the volume flow through a vessel is computed from a through-plane velocity-encoded time-resolved phase-contrast image by summation of the velocities in the vessel multiplied by the pixel area. In principle, time-resolved 3D flow data may be used for the calculation of volume flow by performing this computation on extracted cut-planes from time-resolved 3D flow fields. The volume flow depends on the temporal resolution of the time-resolved 3D flow data, and may be underestimated using this retrospective approach.

Velocity data may be color-coded, in analogy with color-Doppler echocardiography. Since cut-planes can be arbitrarily created in the MRI volume, views that are not easily accessible by echocardiography may be displayed based on the time-resolved three-dimensional MRI data. One way of visualization is to display the flow direction in a slice as vector arrows, either instantaneous, or temporally resolved. This technique usually requires spatial selection of a 2D plane in the data, as vector display of 3D volumes quickly becomes impractical due to the overwhelming number of data points.

5.1.2.1 Particle Trace Visualization

The time-resolved velocity data allows for 3D visualization using particle traces (Buonocore 1998; Wigström et al. 1999; Markl et al. 2011). Particle traces usually originate from an emitting area or volume – even the entire LV volume can be an emitter (Eriksson et al. 2010) – and can be calculated by integrating the velocity field based on the 3D flow acquisition. Different kinds of particle traces such as streamlines and path-lines have been used for visualization of cardiac blood flow using time-resolved 3D phase-contrast MRI. Short instantaneous streamlines generated from a 2D grid display the instantaneous velocity field and thus create an intuitive overview of flow (Fig. 5.1). However, it should be reminded that a blood element will not travel along the streamline over time since a streamline visualizes instantaneous blood flow only.

The motion of blood can be visualized through the trajectories of individual particles. These trajectories, or path-lines, follow the path of virtual mass-less particles released within the time-varying flow field and are used to follow the blood over one or two heart beats as shown in Fig. 5.2. Trajectories are normally computed forwards in time, following an initial condition from a predefined virtual source. Nevertheless, they can also be computed backwards in time, visualizing the origin of the blood flow into to a selected region. Enhanced understanding of the temporal behavior of the path-lines is illustrated by animation, where particle traces are color-coded according to velocity, origin or any other defined parameter. Measurement errors and insufficient correction for background phase offset due to eddy currents and concomitant gradient effects, may decrease the accuracy of particle trace calculations.

Computational errors in the traces may be due to several factors, such as accuracy of the velocity values, quality of interpolation, and difficulty to follow highly curved paths. In particular, such errors accumulate over time, and lead to increasingly inaccurate results. Therefore, care must be taken when evaluating the particle trajectories. The quality of the result must be verified by global conservation rules, such as comparing the number of path-lines entering and leaving a volume (Eriksson et al. 2010). Pathlines are not suited to display mixing within the heart. In fact, the density of path-lines in the different flow regions is highly inhomogeneous, and the

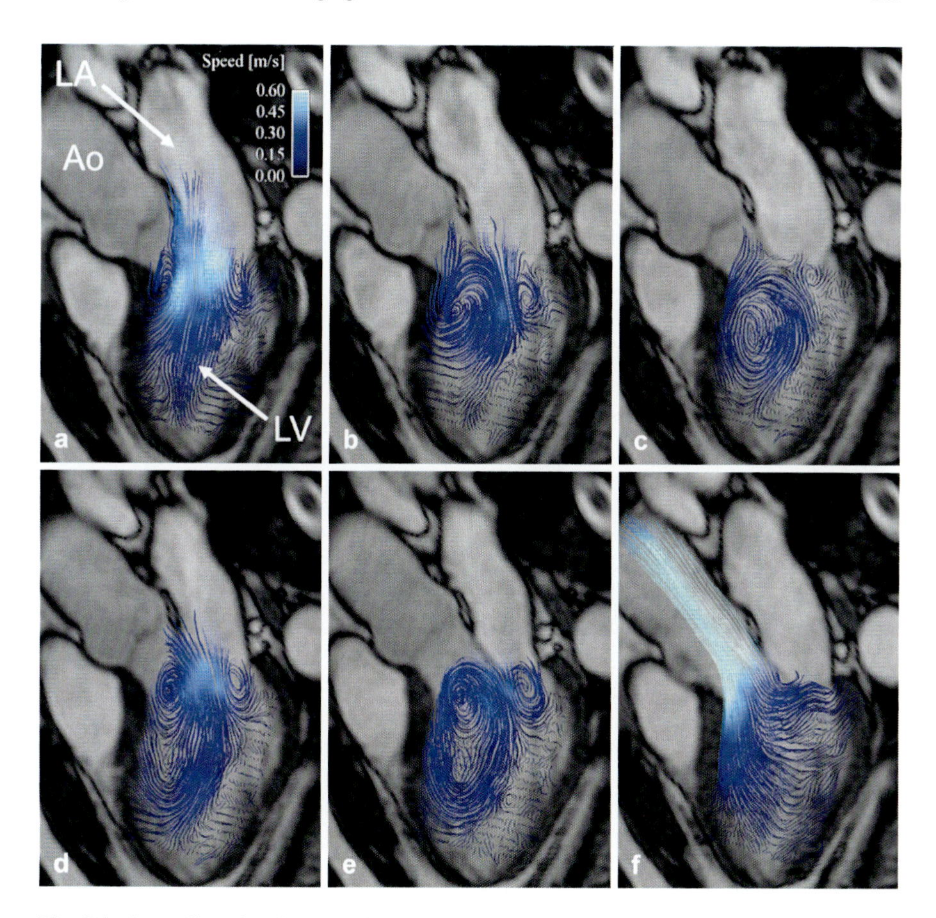

Fig. 5.1 Streamline visualization of blood flow patterns in the left ventricle. Short streamlines were emitted from a grid placed in the long-axis 3 chamber orientation. Some selected timeframes are shown (**a**) peak early diastolic inflow, (**b–c**) diastasis, (**d**) late-diastolic inflow (**e**) pre-systole, (**f**) systolic outflow. *LV* left ventricle, *LA* left atrium, *Ao* Aorta

number of particles entering a given surface with different velocity should be carefully considered to avoid biased results.

With the application of 3D MRI and particle trace visualization, blood flow patterns have been displayed in the entire cardiovascular system (Markl et al. 2011; Ebbers 2011). In the healthy human heart, vortices are demonstrated in diastole in the left ventricle (Fyrenius et al. 2001). Short streamline visualizations are used to study asymmetric redirection of blood flow in the heart, which seems to enhance ventriculo-atrial coupling (Kilner et al. 2000). Large differences in blood flow patterns may be seen in dilated cardiomyopathy compared to healthy hearts (Bolger et al. 2007). 4D phase-contrast MRI has also been used extensively for the assessment of blood flow in the human aorta in health and disease (Markl et al. 2011).

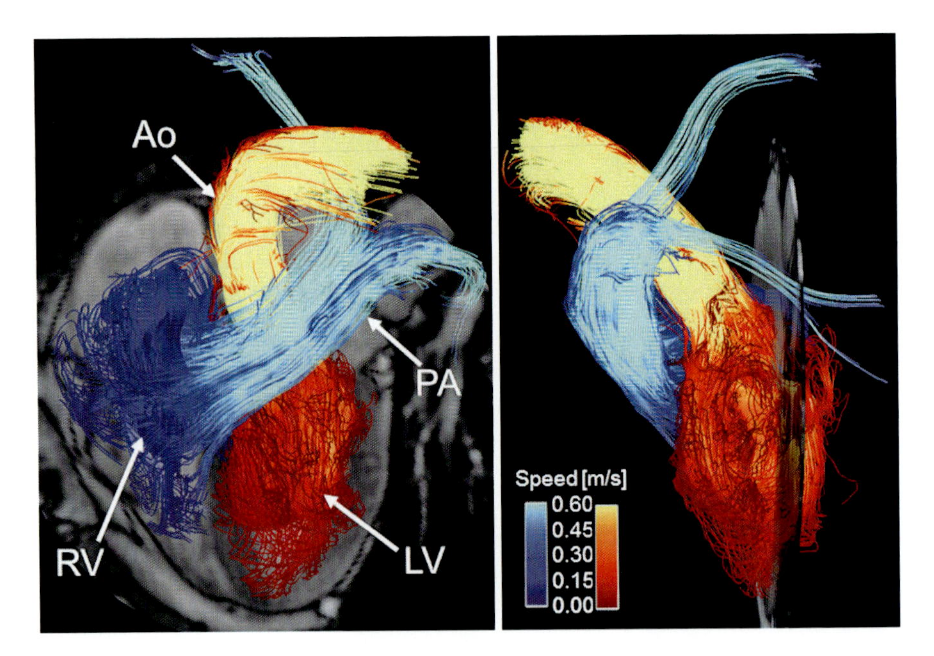

Fig. 5.2 Blood flow patterns in the *left* (*red-yellow*) and *right* (*blue-white*) side of the heart. Pathlines were emitted from the mitral and tricuspid valves at early and late diastole and were traced until end systole. *LV* left ventricle, *RV* right ventricle, *Ao* Aorta, *PA* pulmonary artery

5.1.2.2 Flow Characterization

Currently, the time-resolved 3D velocity data set cannot be visualized in a volume. Therefore, it is usually preferred to automatically extract and only visualize the principal flow structures. Using a pattern matching approach, flow structures such as vortices, swirling flow, diverging or converging flow, and parallel flow can be automatically identified, and visualized (Heiberg et al. 2003), as shown in Fig. 5.3. Visualization can use volume rendering or isosurfaces, or emit particle traces from areas with a high probability of a certain flow structure (Fig. 5.4). Using time-resolved 3D velocity data, the size or volume of the vortex structures is more complex to quantify, as the vortex structures are rather short-lived, asymmetric, and often irregular.

The understanding of cardiac blood flow in healthy and diseased hearts can be extended by assessment of the pressure differences in the cardiac chambers. In general, by assuming blood as an incompressible Newtonian fluid, pressure gradients can be computed from the 4D MRI velocity field using the Navier-Stokes equations. Computational techniques typically utilize the Poisson equation for pressure. This equation is obtained by taking the divergence of the Navier-Stokes equations, under the incompressibility constraint that the velocity has zero divergence. Care must be taken for defining the pressure values at the boundaries when this equation is used. This approach has been used for obtaining the 3D relative pressure field in the aorta (Yang et al. 1996) or the heart (Ebbers et al. 2002). It should be emphasized that only *relative* pressure (compared to the effective pressure value) at an arbitrary point can be recovered from velocity data.

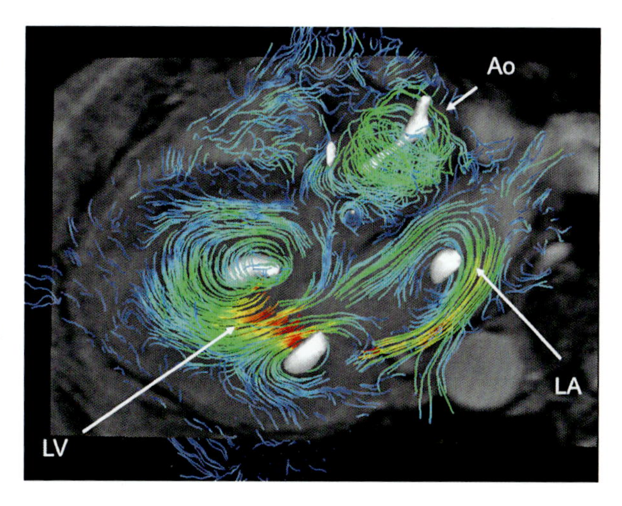

Fig. 5.3 Visualization of the blood flow in the left heart of a patient with dilated cardiomyopathy without valvular regurgitation at late-diastole. The velocity field is visualized using short streamlines generated from a 2D plane combined with blood flow vortex probability (*white isosurfaces*). Notice also the small vortices behind the arotic valve leaflets. *LV* left ventricle, *RV* right ventricle, *Ao* Aorta, *PA* pulmonary artery. Adapted from Ebbers 2001

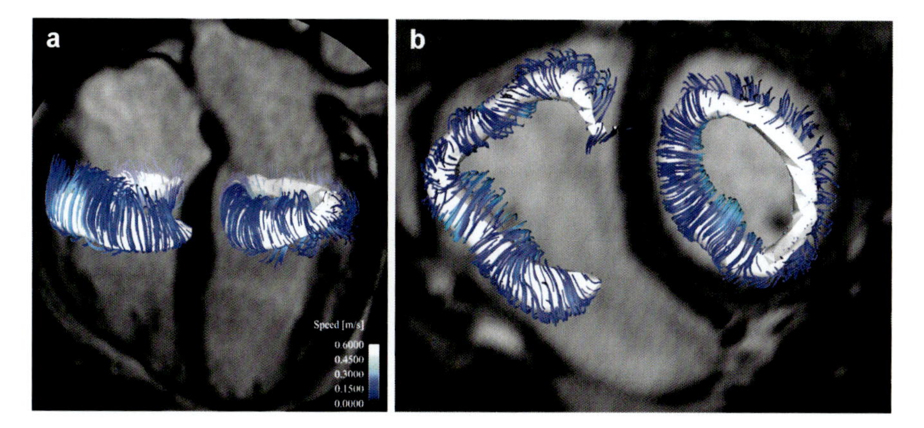

Fig. 5.4 Blood flow patterns in a healthy volunteer during late diastolic inflow in a 4-chamber view (**a**) and viewed from base to apex (**b**). Automatically detected vortex cores are shown as white isosurfaces and streamlines are emitted from these isosurfaces to enhance the visualization. A (partial) vortex ring can be seen below the mitral valve (*right* in each image) and the tricuspid valve (*left* in each image)

5.1.2.3 Quantification of Turbulent Intensity

Disturbed and turbulent blood flow, characterized by fast random temporal and spatial velocity fluctuations, appears to be present in many cardiovascular conditions. These small fluctuations are too fast to be reflected in the standard phase-contrast MRI measurement and diminish the accuracy of the velocity measurement. However, some of these flow aberrations may be used to further characterize the flow properties. Using Fourier velocity encoding MRI, the complete distribution of velocities within a voxel can be measured (Moran 1982), but duration of data acquisition can be relatively long. To this end, the phase-contrast MRI technique has been extended

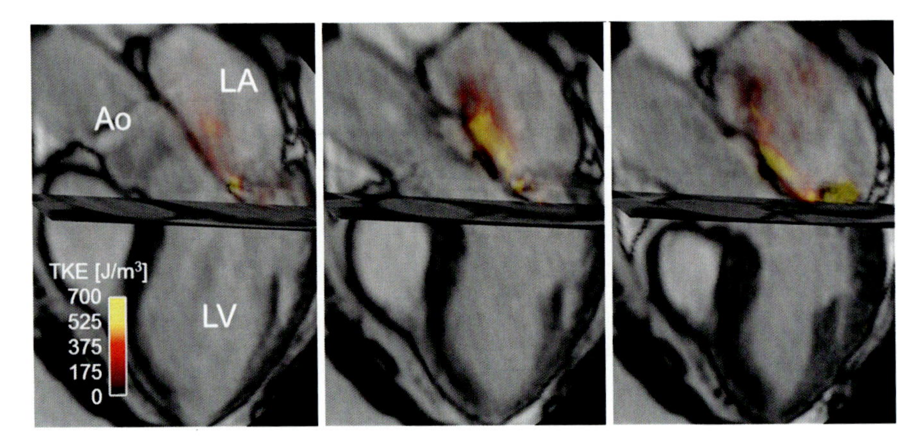

Fig. 5.5 Turbulence intensity: Volume rendering visualization of turbulent kinetic energy (*TKE*) in the left atrium (*LA*) of a patient with an eccentric mitral regurgitation jet due to a prolapse of the middle scallop of the posterior mitral valve leaflet. The left, middle and right panel represent early, peak, and late systole, respectively. *LV* left ventricle, *Ao* Aorta

to the measurement of the standard deviation for a known distribution, which is used to quantify turbulent kinetic energy (TKE) inside a heart chamber (Dyverfeldt et al. 2008), as shown in Fig. 5.5. Principally, both velocity and turbulence intensity can be obtained from the same phase-contrast measurement, allowing for a comprehensive visualization of disturbed blood flow.

5.1.3 Future Developments

Currently, phase-contrast MRI allows an in-depth study of intracardiac blood flow in healthy and diseased conditions (Carlhäll and Bolger 2010; Ebbers 2011; Markl et al. 2011). Phase-contrast MRI is a versatile technique, and future improvements in hardware and software are expected to reduce measurement times. Improved approaches such as merging with fluid dynamics visualization and quantification techniques will also allow the definition and extraction of relevant parameters from these enormous datasets.

5.2 Echocardiography

5.2.1 Blood Flow Visualization Using Echocardiography

Echographic images are generated based on evaluating the returning echoes of previously emitted ultrasound waves, and their delay that depends on the distance of the object from the transducer. The intensity of the echoes allows for the differentiation of tissues based on their echogenicity. Every individual wave is emitted along a one-dimensional scan line. Then several scan lines in sequence are used to span a two-dimensional scan plane or, more recently a three-dimensional volume. Blood itself is echogenic but today's ultrasound systems are not optimized for its visualization, and therefore, dedicated approaches are required to visualize its motion.

The first method for visualizing blood flow is that of injecting an agent that flows with the blood and has better reflectance properties in gray scale imaging. This is typically achieved by infusion of ultrasound contrast agents (perfluorocarbon lipid shell) used for left ventricular opacification, or perfusion imaging, although injected at a lower concentration or solutions based on physiological saline. The second method is based on the Doppler Effect in which the reflected ultrasound wave returns with a modified frequency whose value is proportional to the blood velocity toward or away from the transducer.

5.2.2 Color Doppler

In the early 1990's, Color Doppler technology dramatically enhanced research efforts to visualize and quantitate cardiac flow patterns in normal and diseased hearts. Real time flow patterns on the scan plane were finally being visualized in terms of red and blue colored regions. The Doppler technology has fundamental limitations as it is intrinsically one dimensional, it evaluates the velocity component away and towards the imaging transducer, but gives no information regarding the transversal fluid motion. In other words, it estimates only one component of the three-dimensional velocity vector.

As the velocity is obtained based on the frequency shift of a wave, a positive shift larger than the sampling frequency cannot be separated from a negative shift in the opposite direction. This effect, known as *aliasing*, limits the velocities within a range where the maximum velocity, the Nyquist limit, is given by the frequency half the Pulse Repetition Frequency (PRF). If the Nyquist limit is set too low, high positive velocities will be falsely interpreted as high negative velocities and vice versa. On the contrary, a large Nyquist limit permits to properly evaluate high speeds but reduces the sensitivity to low flow velocities. This is a problem analogous to what described in Sect. 5.1.1.2 for MRI. Over the years, the quality of echographic technology has greatly improved, nevertheless the basic principles of the physics of signal generation and detection remains the same.

Color Doppler imaging provides the ability to qualitatively and quantitatively study the flow and discriminate between normal and abnormal flow patterns. Technology was sufficiently advanced to allow a better understanding of blood motion; it was particularly useful in valvular regurgitation for developing diagnostic measures based on fluid dynamics (Bargiggia et al. 1991; Mele et al. 1995; Grigioni et al. 2003). Changes in flow patterns such as increased velocities through valve orifices, absence of flow with thrombus formation and jets from ventricular septal defects are visualized, resulting in more accurate diagnosis and reduced exam times.

In the mid 1990's, a new technology based on both M-mode and color Doppler technologies was introduced (Stugaard et al. 1994; Takatsuji et al. 1997; Garcia et al. 1998). This technique provided unparalleled temporal resolution with very limited spatial information. A single line of ultrasound data is color coded with velocities to and from the ultrasound transducer. When positioned across the mitral valve orifice the complete inflow patterns from the left atrium to the tip of the left ventricle can be visualized over time. The same can be applied to the left ventricular outflow track that allows visualization of systolic flow with high temporal resolution but limited spatial resolution.

Since flow is related to pressure differences, being able to measure intraventricular pressures has the potential to enhance understanding from cardiac flow dynamics from not only a basic research perspective but also as potential clinical applications. Intra-ventricular pressure gradients were first investigated by the use of invasive pressure transducer catheters. These early studies demonstrated the critical importance of pressure gradients to ensure efficient LV diastolic filling (see, for example, Courtois et al. 1990). Critical differences in the left ventricular flow patterns in various disease and physiological conditions were also identified.

Greenberg et al. (1996) developed a method to estimate the intra-ventricular pressure gradients noninvasively through analyzing intraventricular flow velocities in the left ventricle. By applying basic hydrodynamic principles to the velocity distribution of the left ventricular inflow and integrating the Euler equation along the flow streamlines, they were able to convert velocity information to pressure differences. Although the scan line is not an actual streamline of the flow and the calculations are not rigorous, there was good correlation between the earlier invasive intra-ventricular pressure measurements and the noninvasively generated estimations of left intraventricular pressure gradients (Firstenberg et al. 2000; Tonti et al. 2001).

Numerous clinical studies were conducted to investigate intra-ventricular flow patterns, and specifically the instantaneous pressure gradients inside the left ventricle. Greenberg et al. (2001) focused on the filling dynamics and associated changes with the diastolic dysfunction diseases of LV while Bermejo et al. (2001) centered their study on the outflow gradients and their relationship to ventricular systolic elastance in the assessment of contractility (Fig. 5.6). As with the invasive studies, there were demonstrable differences between various disease states and normal patient populations. Similar to the other methods, this one also suffers from its own limitations. According to the simplified Euler's equation, regional variations in outflow profiles should modify flow velocities, and therefore pressures. However, it is now well-known through experimental models that the flow across the mitral orifice, within the left ventricle and the outflow track is more complex and has a three-dimensional rotational pattern. Therefore, regional differences exist that are not appreciable with color M mode technology. As a result, care must be taken when interpreting the flow velocities and pressures obtained with any technique that inevitably involves simplifying

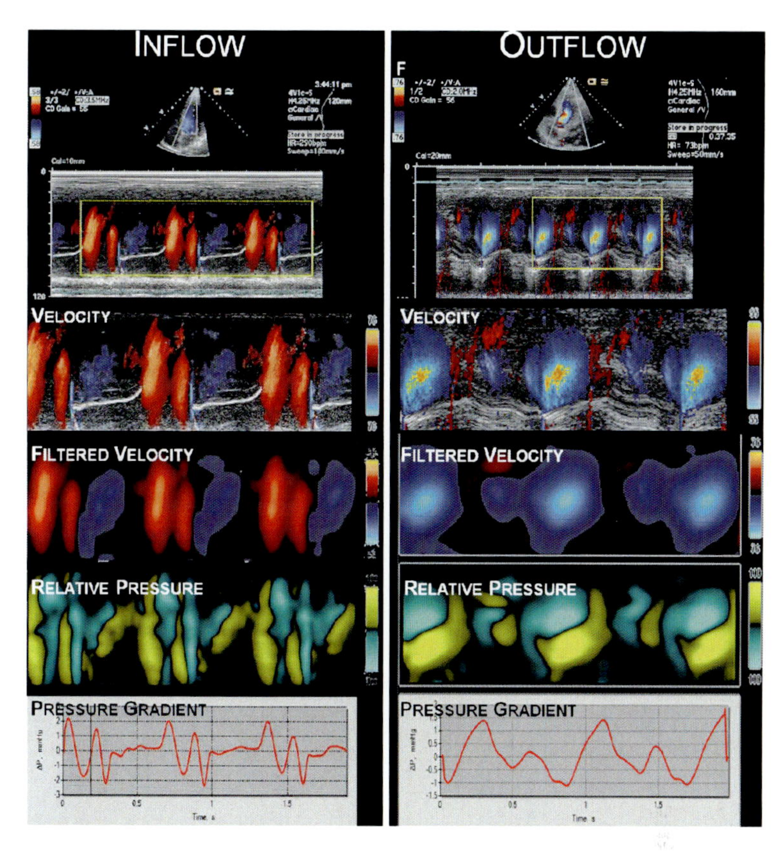

Fig. 5.6 Spatio-temporal mapping of intracardiac pressure gradients for inflow (*left side*) and outflow (*right side*) of the left ventricle. From *top to bottom*: Color Mmode focused on inflow acquisition; Region of Interest focused on three R-R cycles (*Velocity*); Filtered and anti-aliased Color Doppler M-mode (*Filtered Velocity*); Color M-mode velocities converted to Pressure information with Euler's equation (*Relative Pressure*); Intraventricular pressure gradient (IVPG) curve (*Pressure Gradient*)

assumptions. This is even more critical when examining dilated hearts or ventricles with hypertrophy where the rotational flow and vortices are further modified due to LV remodeling.

Currently there are several interesting and innovative approaches to investigate realistic rotational flow patterns within the heart. A recent approach to research is utilizing Color Doppler technology to investigate two-dimensional velocity flow vectors in the heart.

The basic concept in the Velocity Flow Mapping (VFM) is described by Uejima et al. (2010); a method that is based on the same principle then employed by Garcia et al. (2010). These methods assume that the blood motion can be described as a 2D planar flow, initially ignoring any cross flow. Thus, the transversal component of the velocity vector is reconstructed from the Doppler velocity considering the based on the equation. As described in Sect. 1.2 (see Fig. 1.3), every small volume in the cavity must be guaranteed with mass conservation, which results in a relationship between gradients of longitudinal velocity and those of the transversal one. Such a relationship is used to reconstruct the transversal velocity. The VFM technique (Uejima et al. 2010) assumes that blood flow has two components: a single laminar flow and several vortex flow components. The laminar and vortex flow components are separated by use of a simple assumption; the stream function is employed to derive the radial velocity components of vortex flows. In other terms, the flow entering/exiting from the scan plane is assumed to be approximately uniform transversally and balanced by the laminar flow only, while the vortex flow is considered strictly two-dimensional. The technique is being validated utilizing 3D mathematical freestanding and constrained flow models showing that VFM technique may capture the gross features of flow structures and produces images representing the rotational flows in the heart but also overestimates the velocity magnitude of flows (Uejima et al. 2010). Additional studies are required to independently analyze this method. The approach developed by Garcia et al. (2010) is based on a combination of two possible solutions for the transversal velocity that minimizes the difference from a 2D solution.

Either methods still lack of a formulation that permits expression of the type of approximation introduced in the reconstruction. The research is in rapid progress and novel 2D and 3D Doppler imaging techniques based on fluid dynamics concept are currently under investigation.

5.2.3 Contrast Enhanced Imaging and Echo-PIV

The first recorded flow visualization with ultrasound was described in 1984 by Beppu and co-workers in a ground breaking article where they described the distortion to normal flow patterns in ischemic heart disease (Beppu et al. 1988). That article utilized fluoroscopy techniques and applied them to echocardiography. They imaged the normal and ischemic hearts with ultrasound by injecting agitated saline, which acts as a contrast agent, and recorded the flow outlines with video tape recorders. The video clips were systematically reviewed frame-by-frame and the resulting flow patterns examined and described. It was observed that in normal conditions, flow enters the ventricle, reach the apex in one diastolic period, then change direction and move towards the LVOT during systole. In an ischemic animal model, the development of two separate flow streams were identified; one entering the ventricle and immediately turning towards the outflow track, and the other travelling slowly towards the apex along the walls, not necessarily filling the apex during one diastolic period. The degree of observed flow abnormality was directly related to the severity of the infarcted area.

Based on this approach, a method now was developed that utilizes B-mode harmonic imaging seeded with ultrasound micro bubble contrast agents (Fig. 5.7). Such

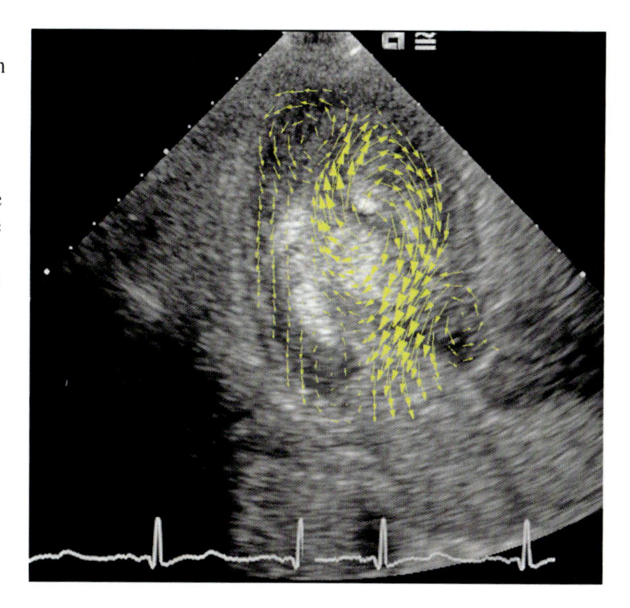

Fig. 5.7 Left ventricular echocardiography acquired in apical long axis orientation with contrast microbubbles, the velocity vector evaluated by echo-PIV are superimposed onto the image to show the flow motion. The vortical flow arrangement at the isovolumetric contraction is shown (Acquisition courtesly provided by G. Tonti)

bubbles visually evidence the swirling intracardiac motion and images can be processed by Particle Image Velocimetry (PIV). This approach allows measurement of the actual two-dimensional velocity field where flow redirection and vortices can be discriminated. The imaging concept benefits from higher image resolution, harmonic imaging and the ability to quantify the flow characteristics. High frame-rate digital movies are analyzed with PIV software developed specifically for echocardiography. Sengupta et al. (2007) and Hong et al. (2008) were the first to report on the advantages of this technique for its ability to differentiate normal and diseased flow patterns.

The echo technique has undergone a careful validation utilizing an *in vitro* model and comparing with laser-based digital particle imaging velocimetry (Kheradvar et al. 2010). The flow in a left ventricle model whose experimental apparatus is described in detail in the next section, was imaged and acquired with high-speed digital camera and high frame-rate ultrasound as shown in Fig. 5.8. The resulting velocity fields and flow patterns compared well. They also evidenced the limitations of the echo-PIV technique. The first is the requirement of very high frame rate (over 200 Hz in normal conditions) otherwise the method presents a cut-off in the velocity magnitude where velocities higher than a prescribed limit (that ranges from 20 to 40 cm/s in normal acquisitions about 70–80 Hz) are underestimated while their direction is still reliable. Additionally, the velocity obtained by echo-PIV has a reduced spatial resolution compared to the original image since a windowing (of size at least 16×16 pixels) is necessary for a statistical estimate of velocities.

Recent works (Faludi et al. 2010) compared valvular normal flow patterns with those after prosthetic valve replacement. The Echo PIV technique revealed presence of a reversed vortical motion in most subjects with prosthetic valve.

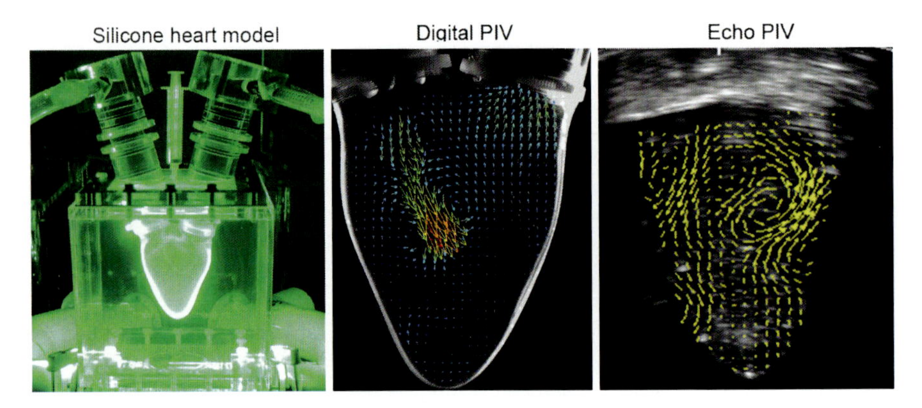

Fig. 5.8 (*Left*) *In vitro* flow phantom attached to pulsatile pump with green laser. (*Center*) PIV processing results of flow phantom from digital camera images. (*Right*) Echo-PIV processing results of flow phantom from ultrasound data

This behavior has been conceptually characterized by a combination of echo-PIV and numerical simulation (Pedrizzetti et al. 2010). The long term outcomes of such flow modifications have yet to be studied but highlight the need for improvement of surgical solutions to maintain normal flow patterns. Further insights into flow patterns of cardiomyopathies, failing ventricles and congenital heart disease are eagerly anticipated.

The echo-PIV technique can be applied, in principle, to 3D imaging. However, the frame rate and spatial resolution requirements limit the translation of the technique in the near future. More promising results are recent advancements based on high frame rate and multiple 2D plane acquisitions that allow reconstructing the principal features of the three-dimensional flow structure (Fig. 5.9).

5.2.4 Future Developments

The field of echocardiography continues to evolve with refinements in existing technologies, new innovations, and applications of advanced analytical algorithms. The technology is well positioned to provide researchers and clinicians sophisticated practical tools to help their patients.

5.3 In Vitro Experiments

This section describes particle image velocimetry technique, which is an important tool to assess cardiovascular flow field. Experiments that characterize vortex formation in the context of cardiac physiology are also discussed.

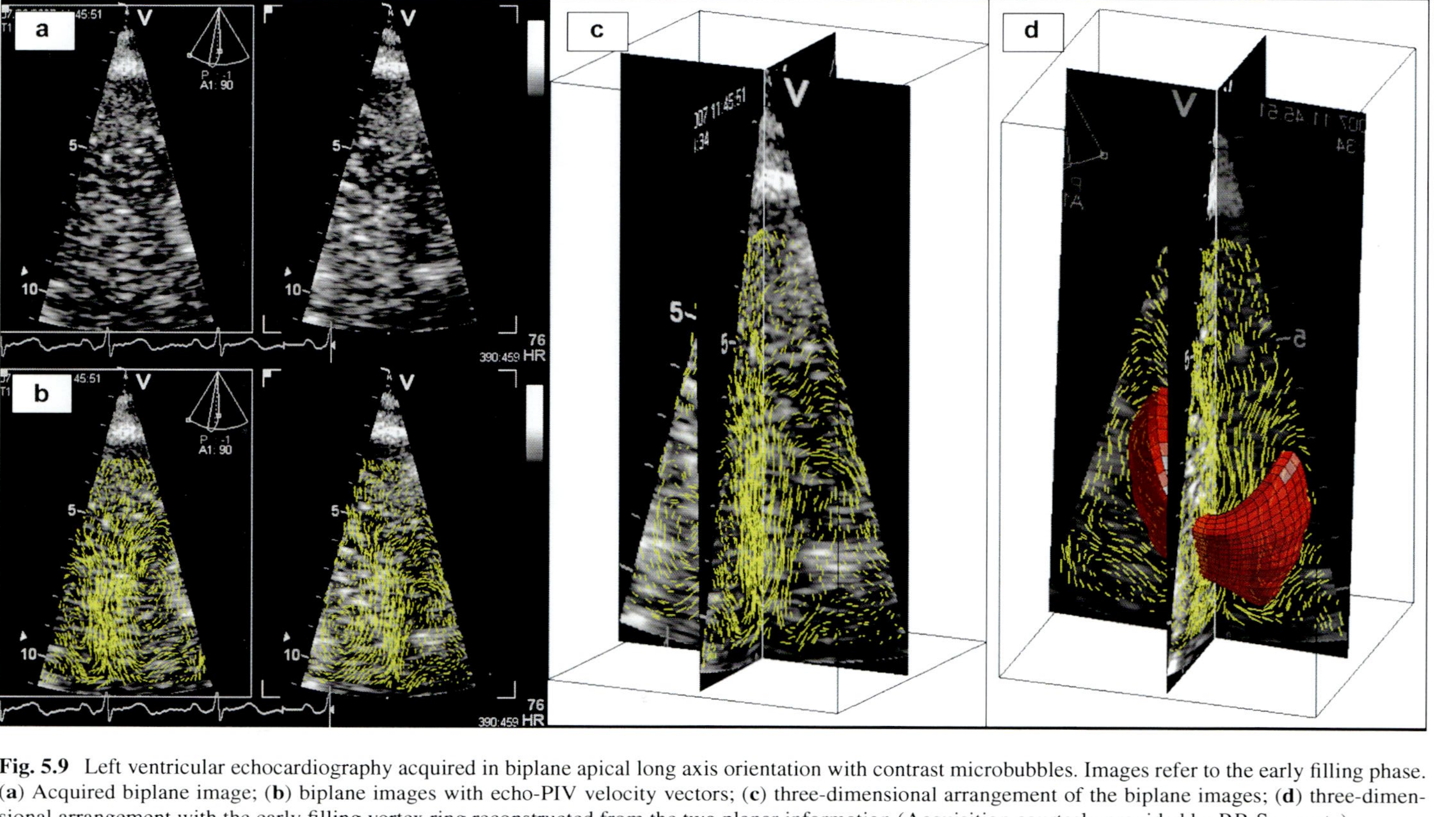

Fig. 5.9 Left ventricular echocardiography acquired in biplane apical long axis orientation with contrast microbubbles. Images refer to the early filling phase. (**a**) Acquired biplane image; (**b**) biplane images with echo-PIV velocity vectors; (**c**) three-dimensional arrangement of the biplane images; (**d**) three-dimensional arrangement with the early filling vortex ring reconstructed from the two planar information (Acquisition courtesly provided by P.P. Sengupta)

5.3.1 Particle Image Velocimetry

Particle image velocimetry (PIV) technique is commonly used to visualize vortex in *in vitro* phantoms related to cardiovascular system. PIV technique was developed in experimental fluid mechanics (Adrian 1991; Willert and Gharib 1991), and its improvements and modifications were then applied to numerous cardiovascular models. PIV is an optical method of flow visualization that obtains instantaneous velocity measurements in a fluid seeded with tracer particles that faithfully follow the flow dynamics (Fig. 5.10). PIV method requires a high-power double-pulsed laser system, a series of optics that generate a condense sheet of laser light and high speed camera(s) synchronized with the laser system to capture the particle field. After the particle fields are captured, the frames are fragmented into interrogation areas through which a displacement vector will be computed based on the motion of the particles. This information is then converted to velocity knowing the time-difference between each particle field.

The time-difference (Δt) between two particle fields is a critical parameter that should be meticulously chosen through experiment based on inherent velocity of the fluid whose velocity field is being characterized. Too large or too small of a

Fig. 5.10 The image represents transmitral vortex formation phenomenon in an *in vitro* transparent model of the left ventricle. The flow seeded by the fluorescent particles is illuminated by a laser sheet passing through the ventricle model. The ventricular sac is a part of the advanced cardiac flow simulator at Kheradvar's lab

time-difference (Δt) will result in significant error in estimation of the velocity. Based on experience, the time-difference (Δt) required for cardiac flow simulation should be in a range of 1 millisecond. A high-speed camera that captures high-resolution images synchronized with a high-speed laser system (~1 kHz) results in ideal velocity fields for heart models at physiological heart cycle ranges (45–120 beat per minute).

It should be mentioned that since the velocity vectors are calculated based on cross-correlation over small areas of the flow field (interrogation window), the obtained velocity field is a "spatially averaged" representation of the actual velocity field.

5.3.2 Heart-Flow Simulator and Reproduction of Cardiac Cycles

Heart pulsed flow simulators are usually made of a compliant heart-shaped structures are connected to reservoirs, which permit filling and emptying similar to a heart chamber. As an example, we will briefly describe a heart pulsed flow simulator with modular elements that replicates several physiologic/pathologic states of cardiac flow fields (Kheradvar and Gharib 2009).

The schematic layout of an exemplary experimental setup is shown in Fig. 5.11. The system is comprised of a thin-wall ventricle, shaped according to molds in the systolic state, and made of transparent Silicone rubber (Fig. 5.10). The ventricular sac is suspended over the Plexiglas atrium (Figs. 5.10 and 5.11) free-floating inside a rigid, water filled, cubic container made of Plexiglas to avoid optical distortion connected to a hydraulic pump system. The Superpump system consists of a piston-in-cylinder pump head driven by a low inertia electric motor. The pump is controlled by a customized MatLab interface (MathWorks, Inc. Natick, MA) compatible with a Data Acquisition (DAQ) device. The interface controls the motion of the pump's piston according to the predefined functions. The periodic, pulsatile flow in the circulatory system is generated as a response of the ventricular sac to the input waveforms provided by the pump. The waveforms are automatically adjusted based on the position, velocity and pressure feedbacks received by the power amplifier. The waveforms induce appropriate suction and forward flow in the Silicone ventricle during a cardiac cycle. Depending on the type of experiment to be performed, artificial heart valves at different sizes can be placed in mitral/tircuspid and aortic/pulmonary positions.

The system allows for several types of waveforms as inputs to imitate different physiological conditions for either left or right ventricles. The waveforms reproduce desired systolic ratios for each experiment. Systolic ratio (SR) is the fraction of time in a cardiac cycle in which the ventricle is in systolic phase. The frequency of cycles can be set to different values for each group of SR ranging from 0.5 to 1.67 Hz (0.5 Hz=30 bpm; 1.0 Hz=60 bpm; 1.2 Hz=72 bpm; and 1.67 Hz =100 bpm) to reproduce operational range for cardiac function. To attain physiological conditions,

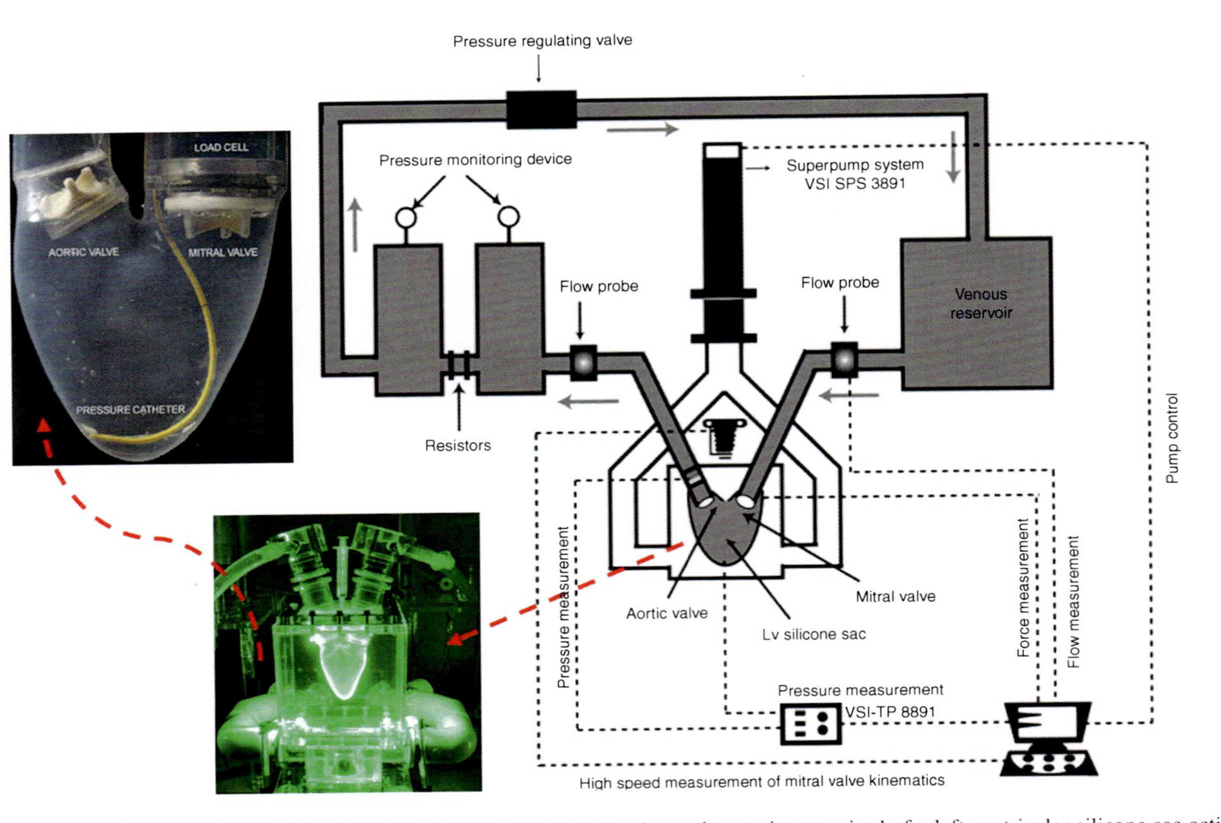

Fig. 5.11 *Right*: Schematics of the heart pulsed flow simulator system. The experimental setup is comprised of a left ventricular silicone sac activated by a suction pump that generates pressure drop outside the sac. High-speed cameras will monitor mitral and aortic valve kinematics while a DAQ system records ventricular pressure, force exerted on the valves and the trans-mitral flow simultaneously. *Left*: Close-up view of the LV sac and its components

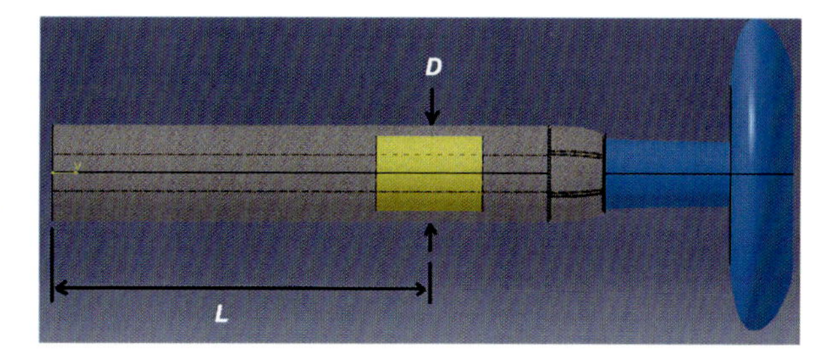

Fig. 5.12 Schematic of a piston-cylinder mechanism for ejection of starting jet and its leading vortex. The length of the jet refers to the distance that the piston travels to develop the jet, and the diameter of the jet is simply the diameter of the piston

the aortic afterload can be set to fluctuate in the range of 80–120 mmHg (mean 100 mmHg) during each cardiac cycle. The pressure at the aortic loop can be measured by a pressure monitoring system during the experiment. The pressure information is used as an input to the feedback control system that adjusts the stroke ratio of the hydraulic pump system.

5.3.3 Experimental Characterization of the Vortex Formation

Vortex rings typically develop from a jet or slug of fluid ejected from a nozzle. In fluid dynamics, both transmitral and trans-aortic flow schemes are considered starting jets ejected from left atrium to the left ventricle and left ventricle to aorta, respectively. The starting jet and its vortex are critically important components of the cardiac cycle, and play multiple roles during systole and diastole. Extensive studies been devoted to the formation and evolution of the leading vortex rings in fluid dynamics. A number of these studies have experimentally addressed stroke ratio of the jet and its correlation with leading vortex. The starting jet is typically produced by a piston-cylinder mechanism (Fig. 5.12), characterized by the stroke ratio as the ratio of the ejected jet length (L) to the effective jet diameter (D). This ratio is often referred to as *vortex formation time* (previously introduced in Sect. 2.4 of this book)

$$VFT = \frac{L}{D} = \frac{V \times t}{D} \tag{5.1}$$

where the length L travelled by the piston is described as the product of its average velocity V and the duration of the pulse t.

Gharib et al. (1998) experimentally demonstrated that by increasing the stroke ratio of a starting jet beyond a critical value, about four in semi-infinite space, no additional energy or circulation enters the leading vortex ring and the remaining fluid in the pulse will be ejected as a trailing jet (see details in Sect. 2.5). After this stage, the vortex ring is said to have pinched-off from the starting jet, and the size of the leading vortex ring does not increase. The critical stroke ratio in which pinch-off occurs is recognized as *vortex formation number*. Rosenfeld et al. (1998) numerically investigated the parameters that affect the starting jets and *vortex formation number*. They described dependence on the velocity profile and the velocity programs in such a way that smoothing the velocity profile decreases the *vortex formation number* and expedite the pinch-off process while smooth acceleration of the piston results in an increase in *vortex formation number* and a delay in pinch-off. In a different experiment using piston cylinder mechanism, Krueger and Gharib (2003) showed that the time-averaged thrust generated by a pulsed-jet in semi-infinite space is maximized once a vortex ring with utmost circulation is created. In other words, once the stroke ratio of a pulsed jet is large enough to pinch off the leading vortex ring and initiate formation of the trailing jet flow, further increase in stroke ratio does not increase the time-averaged thrust. They emphasized that formation of a single leading vortex ring has a more significant role in efficient generation of thrust when compared with the trailing jet.

The left heart acts -in a first approximation- similar to a piston-cylinder system. During the rapid-filling phase of diastole the left ventricle acts as a suction pump that sucks blood from the left atrium as a starting jet. In this phase, the LV can be considered an inverted piston-cylinder mechanism through which fluid is being pulled rather than pushed as in conventional piston-cylinder setups. During diastolic atrial contraction phase, the left atrium acts as a conventional piston-cylinder due to an increase in left atrial pressure that pushes the blood into the LV. In both phases of diastole, the mitral valve is the nozzle that directs the flow. In systole, the left ventricular contraction is the driving force of the flow. During systole, the LV acts as a positive-displacement pump that develops a starting jet passing through the aortic valve as the nozzle.

5.3.4 Influence of Transmitral Vortex Formation on Mitral Annulus Dynamics

The optimization concept discovered for vortex formation process can be seen in a dual perspective in terms of the dynamics of mitral annulus plane in relation to transmitral flow. Kheradvar et al., 2007 used a simplified system (Kheradvar et al. 2007) to experimentally simulate the contribution of the vortex ring on dynamics of the mitral annulus plane during rapid filling phase when the mitral valve is fully open. A vortex ring was generated through a circular annulus with no attached leaflets using exponential pressure drop to mimic the suction effect in the left ventricle

$$P_{LV} = P_0 \, exp\left(\frac{-t}{\tau}\right) \tag{5.2}$$

where P_0 is the left ventricular ambient pressure, τ is the pressure drop time-constant and t is the duration of rapid filling phase. Annulus plane motion and the flow across the annulus were monitored by high-speed cameras during the induced suction. The flow characteristic information (e.g., velocity field and circulation) was captured by phase-averaged digital PIV. Vortex ring circulation during the formation stages was computed from velocity field. Circulation, indicated by the symbol Γ, (defined in Sect. 2.1 and employed in the context of vortex formation in Sect. 2.4); measures the total amount of vorticity entering the vortex core, thus it represents the strength of the forming vortex. Following the vortex formation concept, no energy or circulation is added to the leading vortex when it is pinched-off from the starting.

Through these experiments, it was possible to demonstrate how maximization of the vortex strength presents a second phase of the dynamics of the mitral annulus plane in terms of vortex formation time. Using a physiologic base-apex distance, an annulus diameter of 2.50 cm and a physiologic pressure drop, it was demonstrated that the annulus plane recoil force is maximal once the transmitral vortex ring pinches-off. A similar experiment performed with bioprosthetic valves (Kheradvar and Gharib 2007) allowed for further assessment of the mitral valve dynamics by considering the temporal changes in effective valve opening diameter, which acts as a time dependent nozzle. The results of the experiments with different valve size concluded that the mitral valve recoil is maximal once the vortex ring is about to pinch-off, regardless of the valve size or the characteristics of ventricular pressure drop. Additionally, these results correlated to the onset of the peak annulus recoil with the vortex formation status obtained through simultaneous PIV. It was found that maximal recoil takes place during the period when the vortex is being detached from the transvalvular jet, which occurs at vortex formation time range of 3.5–4.5 (critical vortex formation number range). More information can be found in Fig. 5.13.

These results are attributable to the dilated cardiomyopathy patients. This group of patients may have a significantly larger mitral annulus (Kwan et al. 2003). Therefore, the transmitral formation time is expected to be lowered even if their LV suction is intact (no change in L but larger D). Other researchers (Mori et al. 2004) showed that the peak mitral annulus velocity would also be lower in these patients, which may denote the correlation between transmitral vortex formation and the mitral annulus dynamics.

5.3.5 Conclusive Remarks

Properly designed *in vitro* experiments permit rigorous investigation for characterization of vortex formation process in cardiac chambers. These *in vitro* experiments allow for investigations that cannot be performed through direct *in vivo* observations. The experimentation exhibits basic steps required for the transfer of theoretical considerations to clinical trials.

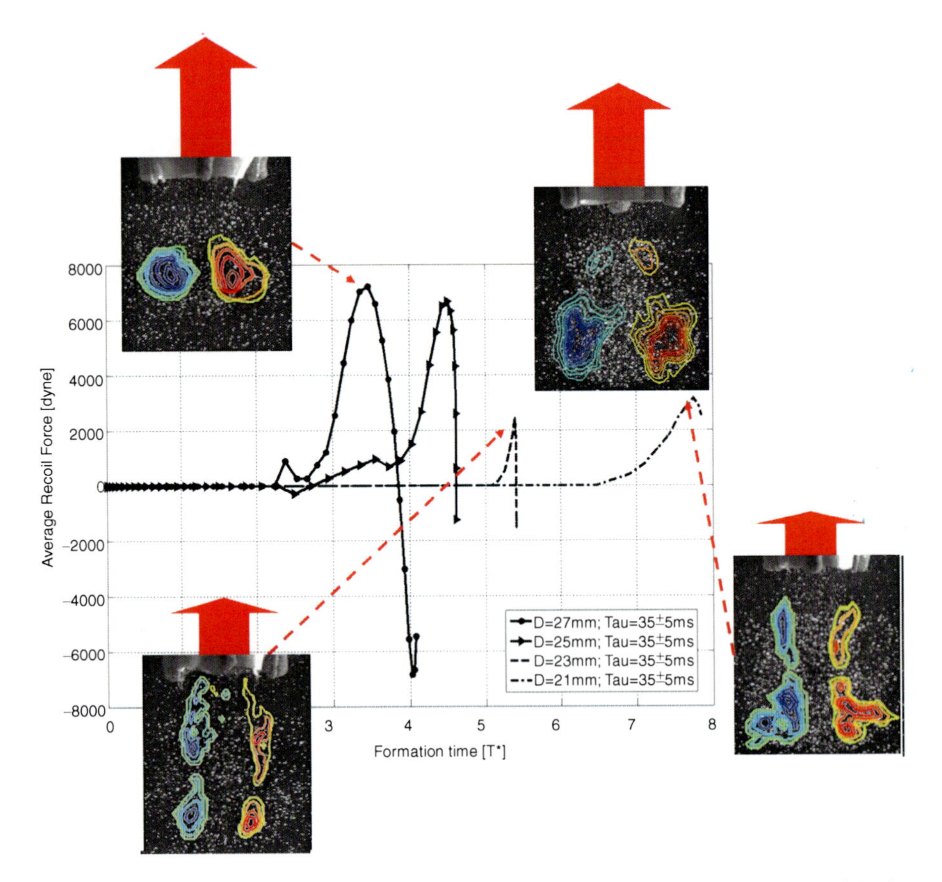

Fig. 5.13 Kheradvar and Gharib measured the annulus plane recoil due to an identical physiologic pressure decay program for different valve sizes simultaneous with valve kinematics and vorticity field. Plotting the annulus plane recoil versus vortex formation time showed that the maximum recoil force occurred during the period that vortex formation time is between 3.5 and 4.5. The vorticity field shows that at vortex formation time 3.5, the vortex ring is very close to pinch off and at vortex formation time 4.5 the ring has just been pinched off. For the smaller size valves, the pick recoil force was found much smaller and the vorticity field show that the vortex ring has already been pinched off and most of the fluid has been ejected as trailing jets

5.4 Numerical Simulation

5.4.1 Fundamentals Elements of Numerical Simulations

By virtue of advances in computational fluid dynamics (CFD), computational structure dynamics (CSD), and rapid growth in computational power, numerical analysis has become a reliable tool for practical engineering purposes. Problems that involve investigation of cardiovascular hemodynamics are frequently characterized with unsteady flows that often take place in complex 3D geometries, which are mostly patient-specific. The flow is often coupled with flexible or passively-moving walls. Therefore, the application of CFD in biological systems still is a challenging task. CFD techniques are based on solving the incompressible continuity (Eq. 1.5) and Navier–Stokes equations (Eq. 1.14) equations for viscous fluid flows. When the numerical simulations do not involve simplification or other approximations from the governing equations, they are called Direct Numerical Simulations (DNS) to specify that the results are theoretically exact.

When accurate measurement of high velocities, or 3D velocities is inaccessible *in vivo,* numerical simulations may serve as powerful tools to complement and provide useful results for evaluation of cardiovascular hemodynamics. With continuing advances in computational power and simulation software along with image-based techniques that provide vascular geometry, flow rates, and velocity profiles, numerical simulation allows good estimates for flow fields to be used in clinical settings. Such information assists in the analysis of cardiovascular conditions, improvement of surgical procedures and developments of cardiovascular implants and assist devices.

Such technological advancements have made it feasible for engineers to provide clinicians with reliable predicting tools, thus continuously reducing the costly and time-consuming *in vivo* animal studies.

5.4.1.1 Discretization Methods

The basic concept in numerical simulations is that the digital representation of a continuous flow field (e.g., velocity or pressure) is given by its value on a discrete set of points evenly distributed on the fluid domain. This set of points is normally referred as the numerical grid, or the mesh, superimposed over the fluid domain. Once the variables are known at the numerical grid points, the derivatives appearing in the continuity and Navier–Stokes equations are replaced by the finite differences between the values of a variable between neighboring points. As a result, the continuity and momentum equations are written on all such discrete points and become a set of algebraic equations for the unknown values of a variable at such points. The solution for such a large system of algebraic equations is a tedious task that is performed by the computer. This numerical method is known as Finite Differences (FD). FD represent the foundation of numerical solutions. They were developed for various technical solutions and specific applications (Moin 2010). Given the simplicity and controllable accuracy of FD methods, they are the preferred choice for theoretical studies and fluid dynamics research. However, they are limited to simple geometries since they are based on the existence of underlying coordinate systems. The FD approach has been used to simulate the 3D flow in a LV using prolate spheroid coordinates (Domenichini et al. 2005).

The majority of studies investigating cardiovascular flow use commercially available CFD software packages where most of their computational approaches are

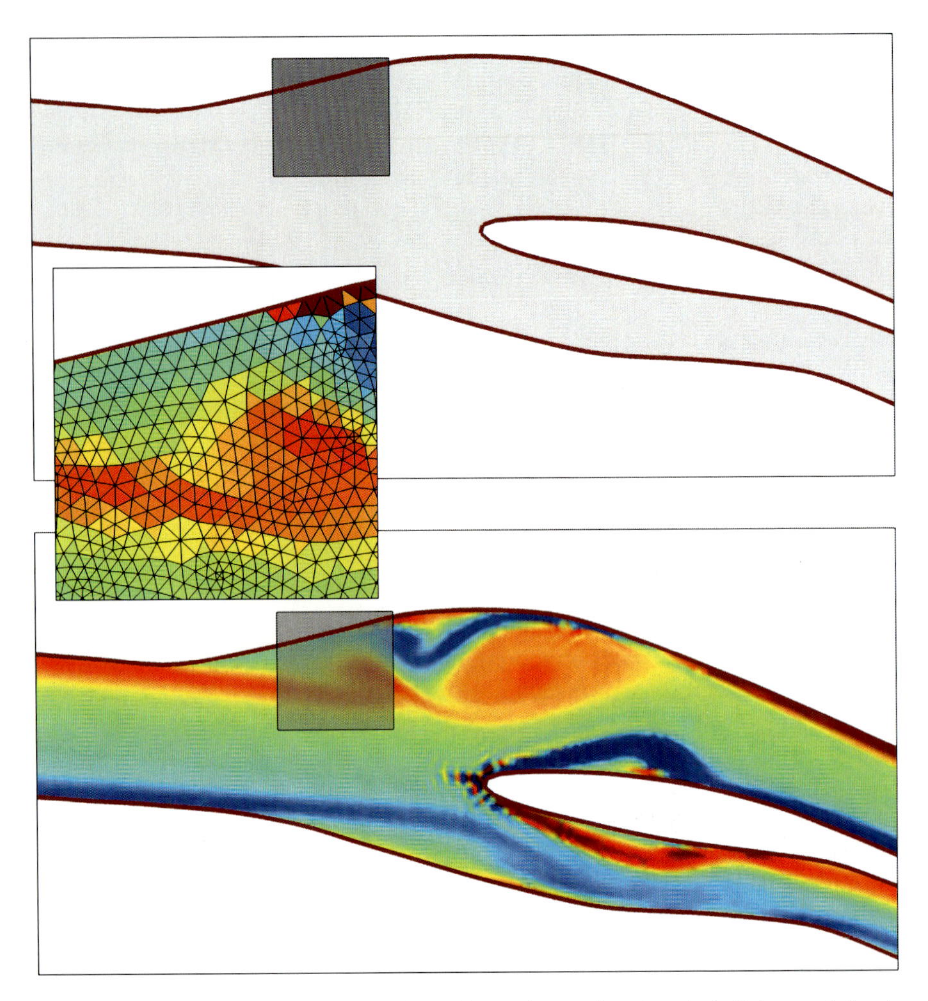

Fig. 5.14 The discretization process tranforms the continuous flow domain, in light gray (*above*), into a large number of triangular elements as shown in the enlarged inset. Each quantity is thus defined inside the individual triangular element (finite element). Eventually, the flow field resulting from the numerical simulations is the visualized as a continuous field (*below*)

based on Finite-Volume Methods (FVM) and Finite Elements Methods (FEM), to convert the integral conservation equations into a combination of algebraic equations. Discretizations in terms of FEM and FVM are known as *unstructured* meshes because they involves subdivision of the flow domain into a large number of finite elements/control volumes that do not necessarily follow a structured 3D coordinate system (Fig. 5.14). FEM use simple functions (e.g., linear or quadratic) to describe the local variations of the flow quantities, such as velocity and pressure, inside the element. For FVM, integrations of the flow over the control volumes and of the fluxes over each volume's boundaries are carried out to meet the conservation equations. These two well established methods (Fletcher 1991) are the most frequently used for various engineering problems, including cardiovascular flows.

Cardiovascular flows are almost always periodic in time, subjected to physiologic pulsatile flow. Therefore, when simulating cardiovascular flow, it is essential to perform time-marching schemes that update the flow field, starting from a given initial condition, and advancing in time for more heartbeats until periodic flow is reached (usually after 3–10 cardiac cycles). Numerical time-marching schemes employ a discretization of the time axis and use the evolutionary (Navier–Stokes) equation to advance a time-step after the other. The time-step must be chosen in agreement with the spatial discretization to ensure that a fluid particle does not travel or diffuses farther than the neighboring cell during one time step. Otherwise, the solution may give rise to loss of stability and numerical results diverging from the actual solution.

However, despite the progress in numerical methods and the continuous increase in power of modern computers, most of the problems involving cardiovascular flow still remain very challenging owing to the complexity of the moving geometries, fluid–structure interaction, intrinsic flow unsteadiness and highly intense velocity gradients in both space and time. Often times several idealizations are required to transform a real cardiovascular phenomenon into a numerically solvable problem.

When the numerical simulation is feasible, the results contain complete detailed information of the flow field, which are available for further processing or for evaluating additional quantities or indicators. Figure 5.15 shows the three-dimensional vortex structure. During the isovolumic contraction, inside a normal left ventricle (left

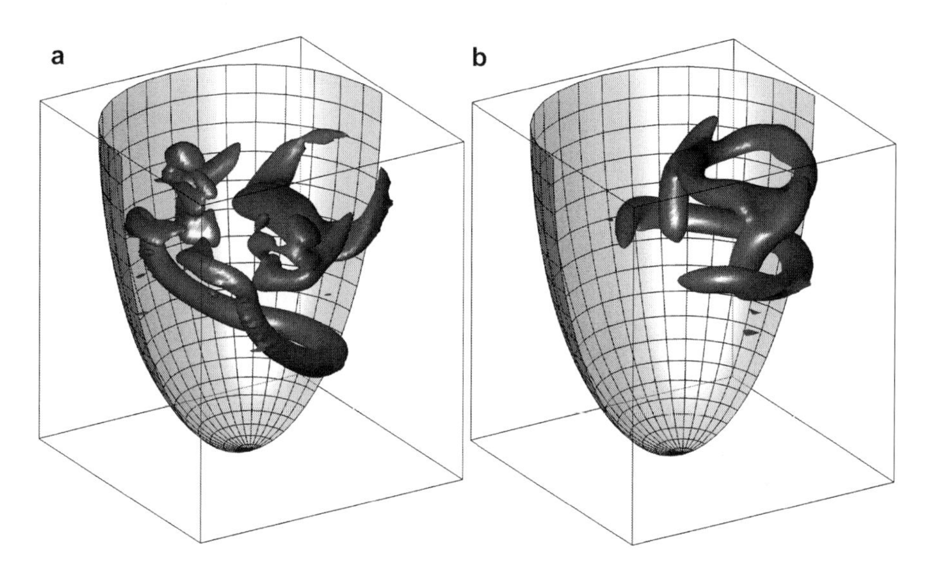

Fig. 5.15 Visualization of the three-dimensional vortex structure at end-diastole in a normal left ventricle EF=55% (**a**), and in a ventricle with apical myocardial ischemia and EF=45% (**b**). The partial occlusion of the anterior descending coronary artery produces an akinesia in the anterior and inferior IVS at median-apical level. Numerical simulations (Domenichini and Pedrizzetti 2011) are performed by the immersed boundary method in Cartesian coordinates. The normal flow shows a vortex structure that reaches the apical region; in presence of mid-apical akinesia the vortex does not enter deeply in the ventricle resulting in a stagnating apical blood pool

side) and in presence of apical ischemia (right side) (Domenichini and Pedrizzetti 2011). The visualization of the structure, by the λ_2 methods (see Sect. 2.5) shows how the pathologic flow cannot be washed out from the apical blood pool.

5.4.1.2 Geometry and Mesh Generation

One technical challenge in developing a numerical simulation is having a proper discretization of the computational domain. Cardiovascular flow develops in complex 3D environment. It is evident that the accuracy of the geometric model and of mesh has a significant influence on the accuracy of the flow prediction by numerical solutions. To simulate the complex patient-specific geometry of the cardiac flow domain, the geometric model is often reconstructed from MRI scans (Saber et al. 2003; Schenkel et al. 2009).

Computational studies that incorporate detailed information about the patient's anatomy and physiology are called *patient-specific*. Their importance lies on the large anatomic variations among patients, and thus significant differences in their hemodynamics and biomechanics. The ultimate goal of these patient-specific simulations is to help assessment of patient-specific risks, test alternative operative plans prior to surgery for a given patient, and outcome planning of certain therapeutic plans. Advances in medical imaging and measurement techniques for cardiovascular function play an important role to develop such models.

One important subject is the automatic generation of a numerical mesh that is appropriate for a given boundary. In FEM and FVM, unstructured meshes with tetrahedral elements may easily fit any geometry and thus can be generated automatically. However, these meshes are considered less accurate and require much more computational resources. Structural or hexahedral meshes are usually preferred. However, they often require various user manipulation of the grid. Recently, some CFD packages offer straightforward generation of hybrid or hexahedral meshes that may accomodate patient-specific simulations.

When the flow is adjacent to a movable domain, the *Arbitrary Lagrangian–Eulerian* (ALE) method is the most preferred technique to allow FEM/FVM continuously follow the deforming boundary. The ALE description combines the advantages of both Lagrangian formulation (where the computational points follow the motion) and Eulerian formulations (where points are fixed in space). In this technique, the nodes of the computational mesh can be moved arbitrarily, allowing significant mesh distortion. The implementation of the ALE technique requires mesh adaptation that assigns mesh-node velocities or displacements at each time-step of the calculation. The remeshing can be performed using mesh regularization algorithms or automatic mesh-adaptation techniques (Donea et al. 2004). Mesh adaptation techniques are usually preferred, since they allow large displacements of the boundaries without the expense of mesh distortion. However, their drawbacks are inaccuracies in the conservation equations during the interpolation between old and new meshes. In fact, any interpolation or regularization method produces an artificial enhancement of viscous dissipation. This effect is often referred to as *numerical viscosity* whose entity can easily become comparable or even larger than the real fluid viscosity. When large displacements exist in the domain, mesh adaptation is highly challenging. In such problems, including the 3D flow across a rapidly moving leaflet of heart valves, frequent remeshings are inevitable (Sotiropoulos and Borazjani 2009).

Recently, *Immersed Boundary* methods (IB), or fictious force method were introduced to reduce the burden of generating complex unstructured meshes matching the segmented geometry. In this approach, the grid of the computational domain is fixed and the walls are considered immersed. This technique was originally proposed in 1982 for fluid-structure interaction in the heart (see Peskin 2002), and then readapted for different applications (Fadlun et al. 2000). The IB method can be optimized by using curvilinear grids that approximately match the overall geometry (Ge and Sotiropoulos 2007) or, in a Cartesian cubic domain where the most efficient computational solutions, such as spectral methods, can be employed (Domenichini 2008). However, regardless of the type of immersed boundary the drawback is in the accuracy of the boundary, conditions.

Mesh resolution is another important concern since the simulation results can be inaccurate where the discrete mesh is insufficient to properly describe the velocity gradients (e.g., in boundary layers, vortex formation, and turbulent regions). Simple estimates can show that good quality simulations in a single cardiovascular structure require a mesh with at least 10^6–10^7 nodes. Computing such fine meshes still may take hours or even days on extremely high power computational platforms.

5.4.1.3 Boundary Conditions

The Navier–Stokes and continuity equations, which are major governing equations of fluid dynamics are second order differential equations. These equations can predict flow characteristics once the surrounding boundaries, known as the boundary conditions are defined. The major differences in flow behavior is due to the types of boundary conditions. Two major types of boundary conditions must be specified; (1) inflow/outflow boundary conditions at the open ends of the computational domain, and (2) boundary conditions at the intersection of the fluid with solid wall.

The inlet and outlet boundary conditions critically affect accuracy of simulation (Groen et al. 2010). The pulsatile flow imposed as the boundary condition is usually a waveform taken from phase-averaged *in vivo* measurements with errors in the range of 5%–20%. In patient-specific simulations, it is critical to impose patient-specific boundary conditions for best correlations (Saber et al. 2003).

To incorporate the distal cardiovascular peripheral resistance, 1-D lumped models or multiscale approaches are frequently used (Formaggia et al. 1999; Migliavacca and Dubini 2005; Avrahami et al. 2006; Diaz-Zuccarini et al. 2008). In these approaches, the boundary conditions are specified through simultaneously solving a simplified model of the entire circulation. This approach has the merit of coupling the local flow structures with the systemic properties. However, it requires defining additional parameters that are normally not accessible in clinical settings.

The boundary conditions at the surrounding structures are conceptually simple when boundaries are rigid or having known velocities. In these conditions, the fluid must exhibit the same velocities as the walls, and the relative velocities are zero. When the underlying computational mesh matches the wall geometry, the boundary conditions are easily specified with relative accuracy. In the most common case of an unstructured grid, it is fitted onto the wall and the boundary condition can be imposed with ease. In the immersed boundary method, when the boundaries do not match with the computational points, the Navier–Stokes equations are solved on the regular domain with the addition of a body force at the boundary position. As a result, the IB method presents a reduced accuracy near the boundary. However, in the context of biological fluid mechanics, this problem may not be critical in most cases, since the structures possess a comparable density to the fluid, and the boundary is physiologically not smooth.

A more challenging problem exists when the boundary condition for the fluid is not explicitly known. Such cases involving *Fluid–Structure Interaction* deserve a special attention and are further described in the following section.

5.4.2 Fluid–Structure Interaction

One of the more challenging problems in CFD is the interaction between fluids and structural components to predict the fast transient response of a compliant structure in a fluid domain. In Fluid–Structure Interaction (FSI) problems, behavior of the structure is dictated by forces of the flow field, and the flow domain depends on the structure location. This makes the problem challenging since the unknown motion of the structure is a boundary condition for the flow and the unknown flow field is a boundary condition for the structure. Problems that involves coupled fluid-structure interactions are quite common in the cardiovascular system. Understanding the complex interaction between the vascular/cardiac wall(s) and blood is vital to understand the physiology of the circulatory system, define the etiology of cardiovascular diseases, and develop prosthetic implants and tissue-engineered substitutes.

In most cases, the overall problem is conceptually divided into two fundamental independent sub-domains; one involving the fluid and the other the solid structure. The strongest coupling between the fluid and the structural domains is achieved when direct (monolithic) FSI methods are used and all fluid and structure governing equations are calculated simultaneously at each time-step. These methods require that both domains being solved using the same method. In addition they require careful coupling formulation of the two domains, which are extremely demanding in terms of computational costs (Pedrizzetti 1998; Bungartz and Schafer 2006; Bluestein et al. 2008; Borazjani et al. 2010).

Weaker coupling approaches use staggered coupling when the fluid and the structure are solved separately and integrated iteratively in turn. The transfer variables between the fluid and structure domains take place at the interfaces on the boundary conditions of both fields. One of the challenges in staggered coupling solutions of FSI with large motions is numerical instabilities that are often treated using special techniques (Avrahami et al. 2006; Nobili et al. 2008). However, the distinction between the two methods is not completely decisive; monolytic techniques often use iterative algorithms solution that require careful handling for convergence, while staggered approaches advance the evolutionary equations with high order implicit schemes that induce a coupling between the domains.

When blood flow interacts with biological structure, (e.g., natural or bioprosthetic valves or vascular wall), the motion of the structural domain is non-linear due to the hyperelastic properties of the biological tissue. Solving such problems requires further computational complexities involving constitutive modeling of the tissue, which is a critical issue in development of patient-specific models.

Mooney–Rivlin models (Humphrey 2002) give reasonably acceptable approximation for the linear behavior of the arterial wall elasticity under small strain conditions, with the exponential term matching the stiffening effect of the material under finite strain conditions. However, elasticity models are for the passive deformation of the tissue compliance and do not consider active contractile properties of tissues such as myocardial walls. Therefore, the FSI modeling of cardiac fluid dynamics need to include a model for the contraction of the myocardial fibers, integrating the immeasurable information with appropriate conceptual schemes (Nordsletten et al. 2011). Such applications are at an early stage and require further advancement.

5.5 Conclusion

Numerical simulations of blood motion in the cardiac chambers and main vessels are currently feasible with good reliability, thus computational models are increasingly replacing *in vitro* models (e.g., *in silico* experiments). Solutions based on FEM or FVM are available in commercial software that provide tools for automatic grid generation and extensive post-processing. The other alternative is the IB methods that entail higher computational efficiency, resulting in future real-time analyses. However their development is at an earlier stage and they are not yet commercially available.

The long term perspective of CFD includes developing patient-specific numerical models by importing medical imaging data as the proper boundaries, performing simulations and post-processing to support clinical decisions.

It should be reminded that numerical models, like every other model, represent limited phenomena within the living systems. Subtle idealization processes from reality may be required as well as integration of missing information with appropriate conceptual schemes. However, oversimplification will always result in erroneous conclusions. Careful interdisciplinary collaborations are highly encouraged regarding application of the computational techniques in clinical practice.

References

Adrian RJ. Particle-image technique for experimental fluid mechanics. Annu Rev Fluid Mech. 1991;23:261–304.

Avrahami I, Rosenfeld M, Raz S, Einav S. Numerical model of flow in a sac-type ventricular assist device. Artif Organs. 2006;30(7):529–38.

Bargiggia GS, Tronconi L, Sahn DJ, et al. A new method for quantitation of mitral regurgitation based on color flow Doppler imaging of flow convergence proximal to regurgitant orifice. Circulation. 1991;84:1481–9.

Bellhouse BJ. Fluid mechanics of a model mitral valve and left ventricle. Cardiovasc Res. 1972;6: 199–210.

Beppu S, Izumi S, Miyatake K, NagataS, Park YD, Sakakibara H, Nimura Y. Abnormal blood pathways in left ventricular cavity in acute myocardial infarcation. Experimental observations with special reference to regional wall motion abnormality and hemostatis. Circulation. 1988;78(1):157–64.

Bermejo J, Antoranz C, Yotti R, Moreno M, Garcia-Fernandez M. Spatio-temporal mapping of intracardiac pressure gradients. A solution to Euler's equation from digital postprocessing of color Doppler M-mode echocardiograms. Ultrasound Med Biol. 2001;27:621–30.

Bernstein MA, Zhou X, Polzin J, King K, Ganin A, Pelc N, et al. Concomitant gradient terms in phase contrast MR: analysis and correction. Magn Reson Med. 1998;39(2):300–8.

Bluestein D, Alemu Y, Avrahami I, Gharib M, Dumont K, Ricotta JJ, et al. Influence of microcalcifications on vulnerable plaque mechanics using FSI modeling. J Biomech. 2008;41(5):1111–8.

Bolger AF, Heiberg E, Karlsson M, Wigström L, Engvall J, Sigfridsson A, et al. Transit of blood flow through the human left ventricle mapped by cardiovascular magnetic resonance. J Cardiovasc Magn Reson. 2007;9(5):741–7.

Borazjani I, Ge L, Sotiropoulos F. High-resolution fluid–structure interac-tion simulations of flow through a bi-leaflet mechanical heart valve in an anatomic aorta. Ann Biomed Eng. 2010;38(2): 326–44.

Bungartz HJ, Schafer M. Fluid-structure interaction: model-ling, simulation, optimisation. Berlin: Springer; 2006.

Buonocore MH. Visualizing blood flow patterns using streamlines, arrows, and particle paths. Magn Reson Med. 1998;40(2):210–26.

Carlhall CJ, Bolger A. Passing strange: flow in the failing ventricle. Circ Heart Fail. 2010;3(2): 326–31.

Cenedese A, Del Prete Z, Miozzi M, Querzoli G. A laboratory investi-gation of the flow in the left ventricle of the human heart with prosthetic, tilting-disk valves. Exp Fluids. 2005;39(2):322–35.

Courtois MA, Kovacs SJ, Ludbrook PA. Physiologic early diastolic intraventricular pressure gradient is lost during acute myocardial ischemia. Circulation. 1990;82:1413–23.

Diaz-Zuccarini V, Hose DR, Lawford PV, Narracott AJ. Mul-tiphysics and multiscale simulation: application to a coupled model of the left ventricle and a mechanical heart valve. Int J Multiscale Comput Eng. 2008;6(1):65–76.

Domenichini F. On the consistency of the direct forcing method in the frac-tional step solution of the Navier–Stokes equations. J Comput Phys. 2008;227(12):6372–84.

Domenichini F, Pedrizzetti G. Intraventricular vortex flow changes in the infarcted left ventricle. Comput Methods Biomech Biomed Engin. 2011;14(1):95–101.

Domenichini F, Pedrizzetti G, Baccani B. Three-dimensional filling flow into a model left ventricle. J Fluid Mech. 2005;539:179–98.

Donea J, Huerta A, Ponthot JP, Rodriguez Ferran A. Arbitrary Lagrangian–Eulerian methods. In: Stein E, de Borst R, Hughes TJR, editors. Encyclo-pedia of computational mechanics, vol. 1. Chichester: Wiley & sons; 2004. Chapter 14.

Dyverfeldt P, Kvitting JPE, Sigfridsson A, Engvall J, Bolger AF, Ebbers T. Assessment of fluctuating velocities in disturbed cardiovascular blood flow: in vivo feasibility of generalized phase-contrast MRI. J Magn Reson Imaging. 2008;28(3):655–63.

Ebbers T. Cardiovascular fluid dynamics - Methods for flow and pressure field analysis from magnetic resonance imaging. PhD Thesis, Linköpings universitet; 2001.

Ebbers T. Flow imaging: cardiac applications of 3D cine phase-contrast MRI. Curr Cardiovasc Imaging Rep. 2011;4:127–33.

Ebbers T, Wigström L, Bolger AF, Wranne B, Karlsson M. Noninvasive measurement of time-varying three-dimensional relative pressure fields within the human heart. J Biomech Eng. 2002;124(3):288–93.

Eriksson J, Carlhall CJ, Dyverfeldt P, Engvall J, Bolger AF, Ebbers T. Semi-automatic quantification of 4D left ventricular blood flow. J Cardiovasc Magn Reson. 2010;12:9.

Fadlun EA, Verzicco R, Orlandi P, Mohd-Yusof J. Combined immersed-boundary finite-difference methods for three-dimensional complex flow sim-ulations. J Comput Phys. 2000;161:35–60.

Faludi R, Szulik M, D'hooge J, Herijgers P, Rademakers F, Pedrizzetti G, et al. Left ventricular flow patterns in healthy subjects and patients with prosthetic mitral valves: an in vivo study using echocardiographic particle image velocimetry. J Thorac Cardiovasc Surg. 2010;139: 1501–10.

Firstenberg MS, Vandervoort PM, Greenberg NL, et al. Noninvasive estimation of transmitral pressure drop across the normal mitral valve in humans: importance of convective and inertial forces during left ventricular filling. J Am Coll Cardiol. 2000;36:1942–9.

Fletcher CAJ. Computational techniques for fluid dynamics, Fun-damental and general techniques, vol. 1. 2nd ed. Berlin/Heidelberg/New York: Springer; 1991.

Formaggia L, Nobile F, Quarteroni A, Veneziani A. Multiscale modelling of the circulatory system: a preliminary analysis. Comput Vis Sci. 1999;2(2–3):75–83.

Fyrenius A, Wigstrom L, Ebbers T, Karlsson M, Engvall J, Bolger AF. Three dimensional flow in the human left atrium. Heart. 2001;86(4):448–55.

Garcia MJ, Thomas JD, Klein AL. New Doppler echocardiographic applications for the study of diastolic function. J Am Coll Cardiol. 1998;32:865–75.

Garcia D, del Álamo JC, Tanné D, Yotti R, Cortina C, Bertrand É, et al. Two-Dimensional intraventricular flow mapping by digital processing conventional color-Doppler echocardiography images. IEEE Trans Med Imaging. 2010;29:1701–13.

Ge L, Sotiropoulos F. A numerical method for solving the 3D un-steady in-compressible Navier-Stokes equations in curvilinear domains with complex im-mersed boundaries. J Comput phys. 2007;225(2):1782–809.

Gharib M, Rambod E, Shariff K. A universal time scale for vortex ring formation. J Fluid Mech. 1998;360:121–40.

Greenberg NL, Vandervoort PM, Thomas JD. Instantaneous diastolic transmitral pressure differences from color Doppler M mode echocardiography. Am J Physiol. 1996;271:H1267–76.

Greenberg NL, Vandervoort PM, Firstenberg MS, Garcia MJ, Thomas JD. Estimation of diastolic intraventricular pressure gradients by Doppler M-mode echocardiography. Am J Physiol. 2001; 280(6):H2507–15.

Grigioni M, Tonti G, Pedrizzetti G, Daniele C, D'Avenio G. In vitro assessment of a new algorithm for quantitative echo measurement of heart valve regurgitant jet. Proc SPIE Int Soc Opt Eng. 2003;5035:287–97.

Groen HC, Simons L, van den Bouwhuijsen QJA, Bosboom EMH, Gijsen FJH, van der Giessen AG, et al. MRI-based quantifica-tion of outflow boundary conditions for computational fluid dynamics of stenosed human carotid arteries. J Biomech. 2010;43(12):2332–8.

Haacke EM, Brown RF, Thompson M, Venkatesan R. Magnetic resonance imaging: physical principles and sequence design. New York: Wiley; 1999.

Heiberg E, Ebbers T, Wigström L, Karlsson M. Three dimensional flow characterization using vector pattern matching. IEEE Trans Vis Comput Graph. 2003;9:313–9.

Hong R, Pedrizzetti G, Tonti G, Li P, Wei Z, Kim JK, et al. Characterization and quantification of vortex flow in the human left ventricle by contrast echocardiography using vector particle image velocimetry. JACC Cardiovasc Imaging. 2008;1:705–17.

Humphrey JD. Cardiovascular solid mechanics: cells, tissues, and organs. New York: Springer; 2002.

Kheradvar A, Gharib M. Influence of ventricular pressure drop on mitral annulus dynamics through the process of vortex ring formation. Ann Biomed Eng. 2007;35(12):2050–64.

Kheradvar A, Gharib M. On mitral valve dynamics and its connection to early diastolic flow. Ann Biomed Eng. 2009;37(1):1–13.

Kheradvar A, Milano M, Gharib M. Correlation between vortex ring formation and mitral annulus dynamics during ventricular rapid filling. ASAIO J. 2007;53(1):8–16.

Kheradvar A, Houle H, Pedrizzetti G, Tonti G, Belcik T, Ashraf M, et al. Echographic particle image velocimetry: a novel technique for quantification of left ventricular blood vorticity pattern. J Am Soc Echocardiogr. 2010;23:86–94.

Kilner PJ, Yang GZ, Wilkes AJ, Mohiaddin RH, Firmin DN, Yacoub MH. Asymmetric redirection of flow through the heart. Nature. 2000;404(6779):759–61.

Krueger PS, Gharib M. The significance of vortex ring formation to the impulse and thrust of a starting jet. Phys Fluids. 2003;15(5):1271–81.

Kwan J, Shiota T, Agler DA, Popovic ZB, Qin JX, Gillinov MA, et al. Geometric differences of the mitral apparatus between ischemic and dilated cardiomyopathy with significant mitral regurgitation: real-time three-dimensional echo-cardiography study. Circulation. 2003;107(8):1135–40.

Lauterbur P. Image formation by induced local interactions. Examples employing nuclearmagnetic resonance. Nature. 1973;242:190–1.

Lee V. Cardiovascular MRI: physical principles to practical protocols. Philadelphia: Lippincott Williams & Wilkins; 2005.

Markl M, Kilner PJ, Ebbers T. Comprehensive 4D velocity mapping of the heart and great vessels by cardiovascular magnetic resonance. J Cardiovasc Magn Reson. 2011;13:7.

Mele D, Vandervoort P, Palacios I, Rivera JM, Dinsmore RE, Schwammenthal E, et al. Proximal jet size by Doppler color flow mapping predicts severity of mitral regurgitation. Circulation. 1995;91:746–54.

Migliavacca F, Dubini G. Computational modeling of vascular anasto-moses. Biomech Model Mechanobiol. 2005;3(4):235–50.

Moin P. Fundamental of engineering numerical analysis. 2nd ed. New York: Cambridge University Press; 2010.

Moran PR. A flow velocity zeugmatographic interlace for NMR imaging in humans. Magn Reson Imaging. 1982;1(4):197–203.

Mori K, Edagawa T, Inoue M, Nii M, Nakagawa R, Takehara Y, et al. Peak negative myocardial velocity gradient and wall-thickening ve-locity during early diastole are noninvasive parameters of left ventricular diastolic function in patients with Duchenne's progressive muscular dys-trophy. J Am Soc Echocardiogr. 2004;17(4):322–9.

Nobili M, Morbiducci U, Ponzini R, Del Gaudio C, Balducci A, Grigioni M, et al. Numerical simulation of the dynamics of a bileaflet prosthetic heart valve using a fluid-structure interaction approach. J Biomech. 2008;41(11):2539–50.

Nordsletten D, McCormick M, Kilner PJ, Hunter P, Kay D, Smith NP. Fluid–solid coupling for the investigation of diastolic and systolic human left ventricular function. Int J Numer Methods Biomed Eng. 2011;27. doi:10.1002/cnm.1405.

Pedrizzetti G. Fluid flow in a tube with an elastic membrane insertion. J Fluid Mech. 1998; 375:39–64.

Pedrizzetti G, Domenichini F, Tonti G. On the left ventricular vortex reversal after mitral valve replacement. Ann Biomed Eng. 2010;38:769–73.

Pelc NJ, Herfkens R, Shimakawa A, Enzmann D. Phase contrast cine magnetic resonance imaging. Magn Reson Q. 1991;7(4):229–54.

Peskin CS. The immersed boundary method. Acta Numerica. 2002;11:479–517. doi:10.1017/S0962492902000077.

Pierrakos O, Vlachos P, Telionis D. Time resolved DPIV analysis of vortex dynamics in a left ventricular model through bileaflet mechanical and porcine heart valve prostheses. J Biomech Eng. 2004;126:714–26.

Querzoli G, Fortini S, Cenedese A. Effect of the prosthetic mitral valve on vor-tex dynamics and turbulence of the left ventricular flow. Phys Fluids. 2010;22:041901.

Rosenfeld M, Rambod E, Gharib M. Circulation and formation number of laminar vortex rings. J Fluid Mech. 1998;376:297–318.

Saber NR, Wood NB, Gosman AD, Merrifield RD, Yang GZ, Charrier CL, et al. Progress towards patient-specific computational flow modeling of the left heart via combination of magnetic resonance imaging with computational fluid dynamics. Ann Biomed Eng. 2003;31(1): 42–52.

Schenkel T, Malve M, Reik M, Markl M, Jung B, Oertel H. MRI-based CFD analysis of flow in a human left ventricle: methodology and application to a healthy heart. Ann Biomed Eng. 2009;37:503–15.

Sengupta PP, Khandheria BK, Korinek J, Jahangir A, Yoshifuku S, Milosevc I, et al. Left ventricular isovolumic flow sequence during sinus and paced rhythms: new insights from use of high-resolution Doppler and ultrasonic digital particle imaging velocimetry. J Am Coll Cardiol. 2007;49:899–908.

Sotiropoulos F, Borazjani I. A review of state-of-the-art numerical methods for simulating flow through mechanical heart valves. Med Biol Eng Comput. 2009;47(3):245–56.

Steen T, Steen S. Filling of a model left ventricle studied by colour M mode Doppler. Cardiovasc Res. 1994;28:1821–7.

Stugaard M, Risoe C, Ihlen H, Smiseth OA. Intracavitary filling pattern in the failing left ventricle assessed by color M-mode Doppler echocardiography. J Am Coll Cardiol. 1994;24:663–70.

Takatsuji H, Mikami T, Urasawa K, Teranishi JI, Onozuka H, Takagi C, et al. A new approach for evaluation of left ventricular diastolic function: spatial and temporal analysis of ventricular filling flow propagation by color M-Mode Doppler echocardiography. J Am Coll Cardiol. 1997;27:365–71.

Tonti G, Pedrizzetti G, Trambaiolo P, Salustri A. Space and time dependency of inertial and convective contribution to the transmitral pressure drop during ventricular filling. J Am Coll Cardiol. 2001;38:290–1.

Uejima T, Koike A, Sawada H, Aizawa T, Ohtsuki S, Tanaka M, et al. A new echocardiographic method for identifying vortex flow in the left ventricle: numerical validation. Ultrasound Med Biol. 2010;36:772–88.

Wigström L, Ebbers T, Fyrenius A, Karlsson M, Engvall J, Wranne B, et al. Particle trace visualization of intracardiac flow using time-resolved 3D phase contrast MRI. Magn Reson Med. 1999;41(4):793–9.

Willert CE, Gharib M. Digital particle image velocimetry. Exp Fluids. 1991;10:181–93.

Yang GZ, Kilner PJ, Wood NB, Underwood SR, Firmin DN. Computation of flow pressure fields from magnetic resonance velocity mapping. Magn Reson Med. 1996;36(4):520–6.

Index

A. Kheradvar, G. Pedrizzetti, *Vortex Formation in the Cardiovascular System*,
DOI 10.1007/978-1-4471-2288-3,

Printed by Publishers' Graphics LLC
SO20120728